I0519070

It Never Snows in Vietnam

BYRON OPENDACK

IT NEVER SNOWS IN VIETNAM

Copyright © 2023 Byron Opendack.

All rights reserved. This book may not be reproduced, transmitted, or stored in whole or in part by any means, including graphic, electronic, or mechanical without the express written consent of the publisher except in the case of brief quotations embodied in critical articles and reviews.

This is a work of fiction. The events and characters described herein are imaginary and are not intended to refer to specific places or living persons. The opinions expressed in this manuscript are solely the opinions of the author and do not represent the opinions or thoughts of the publisher. The author has represented and warranted full ownership and/or legal right to publish all the materials in this book.

ISBN: 978-1-962363-45-7 (sc)
ISBN: 978-1-962363-46-4 (e)

Author's publication depository: Eastern Washington University, John F. Kennedy Library, Archives and Special Collections. Accession No. 88-489

Rev. date: 12/18/2023

PROLOGUE

When I was young, my toys were not, but store-bought joys are soon forgot, and as I rocked in Jeffery's chair I sensed his loss and deep despair. Caught up in dreams of years long past that led to horrors none forecast, I see within my young mind's eye: my life, my home and apple-pie.

With night clothes donned and head laid back I rocked to the dipsomaniac who crooned, "Memories are made of this,"—these reveries, my nemesis. Sounds drift up from a room downstairs where plates of spite fly through the air and smash upon the floor of dreams where insects crawl and reign supreme.

When he was four or five, I think, when Dad was young and Mom wore pink, they went back East and left Jeff home—they said they'd write, they'd telephone. But, when my parent's backs were turned, at Christmas time, that's when they learned that Jeff was gone, taken away by his real dad one winter's day.

It was too late when they returned from ill-spent days, pointless sojourn, to stop the wheel's that fate had spun—Jeff's lonely life had just begun. Everything was spinning around, engulfing him, and pulling him down. A little boy can't understand why Mommy left her little man.

As I grew up I learned the truth—how Jeffery spent his damaged youth— standing by the road all day, looking off and far away. Every day he took his stance and gave young life another chance. He waited patiently all day. He never left to sing or play.

1

Aunt Irma said when she drove by she stopped to ask my brother why he kept on looking up that hill. Why did he wait? Why this vigil? "I'm waiting for my mom," he said. And, she could see his eyes were red. "I'm sure she's bound to be here soon. I've been so good—and cleaned my room."

A little boy grew up, it seems, and spent his life exploring dreams with half his soul left in the snow, a gentle man no one will know. Whisked far off when he was young, Jeffery's song was never sung— unless the discharge of a gun can take its place and count as one.

Snow now falls upon his grave and life no longer keeps him slave to memories he can't erase—of open arms, but no embrace. Sleep my brother, don't you cry, nor waste your spirit asking why warmth found you not, nor touched your heart, but kept you and your mom apart.

—Benny Olstein

Traveling at 600 miles a second, the round fired from a Kalashnikov AK47 slammed into his head, tore through his brain and burst out the other side, taking part of his skull and blonde hair with it.

He'd been hit.

There was an explosion and the world went white in a blinding flash. Everything stopped. He felt at peace and imagined himself smiling as he floated through a gray fog. Then, after a time, he was able to open his blue eyes again.

Jeffery wasn't sure where he was. He wasn't even sure he'd been hit. All he knew was one minute he was fine and the next he was lying on his back. A steady drizzle of water made its way over the edges of leaves overhead, suggesting it had rained. He tried to think back, but couldn't remember it raining, not recently anyway.

Sunlight streamed through the jungle's awning while moisture dripped from the towering trees. He listened, but he heard nothing. Once in a while he thought he heard something, but it was only the thumping of his own heart.

When he realized that, he felt himself smile because he was pretty sure he'd cheated death one more time. At least he thought he smiled. In reality the muscles on his face were frozen, he couldn't smile—he couldn't move at all. His lower extremities shook and convulsed uncontrollably, but those were just his muscles and nerves giving up the ghost.

His eyes were still operating; he was able to blink, but even that seemed fleeting. At least he was alive, he shouted to himself, and goddamn it—that was something!

Then, like a dying campfire, the light began to fade and Jeffery felt himself blacking out.

The soldiers who survived the attack bolted for the dense foliage and hit the dirt. The air was heavy, rich with moist heat and putrid from the stink of rotting vegetation. Somewhere above, hidden by the jungle's canopy, the thumping cadence of rotor blades grew louder.

"Jeff's been hit!" Bo shouted, attempting to rise.

"Stay put, asshole. He's not going anywhere. Wait for the choppers," Sgt. Black advised.

There was nothing like the sound of approaching choppers when it was time to go home. It was truly sublime watching those giant metal insects draw near.

The sun bounced from the skids—mosquito legs of steel—as the choppers hovered then slowly sank to the earth. A swarm of blades chopped the air in deafening whacks as great clouds of dust and dirt launched from the downwash.

They had arrived—their taxis home. Those great green wasps, wet with venom; the dragonflies from heaven—their ambulance and their hearse—their lifeline and their deathbed—all rolled into one.

The world had gone black. His eyes had closed at the end of a blink and remained so. But he wasn't dead—not yet, anyway. He was conscious, now. He sure as hell couldn't see, and he couldn't move either, but at least he wasn't in any pain and he had some semblance of awareness. He could be thankful for that; at least he wasn't a vegetable.

Music. Music? Why did he hear music? It was drawing nearer, growing louder, the guitars rumbling, trying to force their way through a filter of chopping rotor blades.

They say the last thing to go is your sense of hearing when you're dying. At least he could still hear, barely. He wasn't sure, but he thought he heard explosions. They must have been close because of their intense heat, but to Jeffery the sound was severely muffled. Sometimes he thought he was on fire and tried to smack the flames by flapping his arms, but he never really moved anything. Chopper blades thrashed, whipping the tall grass into a frenzy, beating against his sides and slapping his face.

He wasn't sure, but he thought perhaps he was floating. No, not floating— nobody floats. Lifted. Yes, that's it, he was being lifted. But, it kind of felt like he was floating. It was almost like flying because it was happening so fast. His buddies were carrying him. He was soaring! Nope, now he was landing—gently, too. Thanks, pals—whoever you are.

"You're goin' to be alright," Bo assured him. "Yes sir, this our ticket home, buddy. I'll be sittin' right here, next to you." Bo patted Jeff's hand.

The chopper hesitated for a moment, as if it wasn't sure, then slowly began to lift off. Movement. He felt himself flying again! There was another explosion nearby, but the music continued.

It was the voices of women—and they were singing. "Soldier Boy." Jeff wanted to sing along, but that was impossible so instead he listened to the words of The Shirelles and tried to remember when he'd first heard that song. But it was so long ago and so far away.

One thing led to another, and he began to remember other times and other places, when the earth was cool and the nights were quiet. He felt himself being swallowed by those memories and drifted off to . . .

. . . that morning when he walked his cousin Nellie to Hangman's Creek. He remembered lying on his back with his arms crooked behind his head studying her bare naked bottom and polka dot dress. She straddled the knot on the end of the giant rope and swung out over the babbling water. Man that was a long time ago, Jeffery thought. It almost didn't seem real anymore.

The tired spring complained before the screen door slammed and Jeffery bounded down the back steps, heading for the barn. He shoved his hands into his pockets, hunched his shoulders against the morning chill, and kicked the cow turds that lay in the dusty drive. The driveway wound past the old chicken coop with its graying wood and collapsing roof—the loose dirt turning to white gravel as he made his way around his aunt and uncle's house.

Summers in eastern Washington were hot and dry. In the mornings the bees took wing with the rising mists to inventory the sleeping blossoms whose petals had yet to unfold. On Aunt Irma's farm the songs of the morning birds harmonized with their two roosters, Barney and Peabrain, as a breeze rustled the dense foliage of the giant poplars that blocked the heavy winds that sometimes swept down from the north.

His Uncle Harold peeked through the unfinished window-frame of the new hen house and watched Jeffery as he made his way to the barn. Duke, Uncle Harold's dog, stopped panting long enough to retrieve a hammer. Lifting his head high, he grabbed hold of the handle with his jaws and pranced about, immensely pleased with himself.

Barney and Peabrain, the two roosters, had exhausted themselves from crowing in the new morning sun and were now busy trying to kill each other. They chased about the barnyard stopping occasionally to engage in a round of fighting. Feathers hung in the air like weightless snowflakes. Jeffery barely noticed.

"I wonder what's got into that boy," Harold mused aloud. Duke thought Harold was talking to him and stopped to listen. "He usually helps us set up before breakfast."

Duke thought Harold's words required a response and so relaxed his jaw enough to let the hammer tumble out of his mouth onto Harold's foot.

"Jesus!" Harold shrieked. The pain shot up his leg and homed in for his lumbars. "Lord have mercy. Watch what the hell you're doing, Duke. Damn!" Duke grinned while panting, and Harold limped around in a circle, waiting for the pain to subside and wondered if Irma had heard him. He could swear all he wanted, but Irma would not tolerate taking the Lord's name in vain. Sometimes it just slipped out anyway.

Harold nudged his spectacles and peeked over the rims; no movement from the house, Irma hadn't heard. He turned his attention to Jeffery. He was walking away, deep in thought. "Guess it's just you and me this morning, boy. Jeffery looks like maybe he's got other plans."

Duke barked in agreement and panted a few times. Saliva drooled from the corners of his mouth and made tiny splashes where it landed on the new floor.

Harold resumed work anew—limping, as he adjusted the saw horses, plugged in the power tools and admired the work completed the day before. He bent over and retrieved the hammer. "Damn it, Duke. You got the handle all wet with slobber." Duke wasn't paying any attention—he had decided a tarp needed moving and was attempting to accomplish the job, tugging at it. Harold looked out the window once more.

Jeffery shuffled into the barn, nearly stepping on a giant grasshopper that rocketed suddenly—sailing over his head, its wings clapping loudly as it fled for the safety of the alfalfa field.

It was cool inside the barn and a little dark at first. The barn's ceiling was so high the wooden beams were obscured by a dust cloud that hovered near the top loft. When Jeffery grabbed hold of the old wooden ladder and looked up, grime and hay dust broke loose from the haze and fell into his eyes, stinging like a dry rain of nettles.

For Jeffery, barns were shelters from the emotional storms of adolescence; they were warm beds—mountains of ambrosial hay that poked him in the neck and hugged him with yellow bricks of neatly stacked bales; they were stables, and harnesses and abandoned farm machinery that sat for years in the same place and rusted—being eaten by the elements while providing shelter for wild buttercups and field mice.

Jeffery climbed to the highest loft where the ceiling beams supported a roof that hugged the sky. He left the ladder and found a spot in the hay where the rising sun poured through the missing slats in the walls of the barn.

Hard to believe it was only yesterday when he last saw Nellie. Beneath that polka dot dress Nellie had been as naked as a Jay bird.

She never wore any underwear and bragged about it, too. Straddling that big old knot on the rope swing, her dress billowed out as she sailed over Hangman's Creek. It was all Jeffery could do to concentrate on what she was saying as he followed her bare bottom over the rushing water.

More important than his cousin's naked butt had been her words, he couldn't get them out of his head. If what she'd said was true, he needed a lot of time to think.

She was constantly jabbering—talking on and on. Most of the time her words were just that, noise. But this time he'd listened—and this time even Nellie's bare butt couldn't distract him from their importance.

The summer of 1952 was one of those times when Jeffery would stay at his aunt and uncle's farm for an extended period of time. His home life was less than comfortable; his father seldom engaged him in conversation and his Aunt Jackie, whom his father lived with, seemed strangely preoccupied with her stupid books and magazines.

Anyway, he liked staying at Irma and Harold's. They were both nice enough and anyone could tell they were genuinely in love. In fact, he supposed if he had to choose anybody he'd rather spend his summer's with, it would probably be them.

Irma was short and had more than her fair share of wrinkles, but her manner made her attractive just the same. It seemed most of his cousins were more than comfortable around her, in fact they were infatuated with her and would spend hours in her kitchen helping with the cooking chores.

Their large white farmhouse stood on a hill-crest nestled between two ancient maple trees. In the rear, where the manicured lawn dipped until it reached the edge of the woods, one maple tree leaned ominously over the back porch.

"I'm gonna have to cut that son of a bitch down one of these days," Harold threatened. In the evenings when the crickets sang, Irma and Harold sat on the porch swing and rocked to the rhythm of the night air.

Jeffery's room, when he stayed with his Uncle Harold and Aunt Irma sat directly over the back porch. When the summer moon was high and the bats took to the air to peruse the countryside, Jeffery lay awake listening to the creaking of the porch swing and the soothing voices of friendly bickering.

"What'd that tree ever do to you?" Irma asked, returning from the kitchen with their second glass of iced tea. "I love that old tree."

"I do too," said Harold, accepting the glass of tea. "But I don't like the way it's leaning."

To Jeffery's ears, out of all his uncles, Harold possessed the most distinctive and likable of voices. Shilly-shallying to the likes of Andy Divine; the octave pops occurring unexpectedly—surprising both the speaker and the listener.

"I'm telling you that damn tree is dangerous. It wouldn't take much of a storm to uproot it and send it flying through the damned house—probably killing both of us. Nope," said Harold chugging his tea, "I'm afraid that sooner or later I'm gonna have to chop her down. Ain't that right, Duke?"

Duke's ears perked up and he barked in response. The dog had been napping, stretched out on the top step. At the mention of his name he sat up and pounded the deck with his tail.

"See, Duke agrees."

Duke panted a few times, swallowed a mouthful of saliva, then slumped back down, resting his large black head on his front paws before closing his eyes.

"I think you've had too much iced tea and it's gone to your brain," said Irma rising. "Come on, let's go to bed, it's getting late and I've got to get ready for the family picnic."

"All right," said Harold as he forced himself out of the porch swing. "Duke?" Duke's head jerked up and he looked at Harold.

"You're in charge tonight. Keep an eye on the farm while we're sleeping, ya hear?"

Duke barked in response, smiled, swallowed a mouthful of saliva and went back to sleep.

Duke was Harold's dog. Irma fed him, bathed him when he needed it, and smacked him when he needed that, too—-but Duke was devoted to Harold. Eight years earlier, Duke showed up on the back porch during one of the worst rain storms in a decade. Irma was closing up the house for the evening when she heard a scratch at the door and the whimper of a puppy.

"What the hell is that?" Harold asked, joining her in the doorway. The mutt snuggled closer to the screen door to avoid the rain.

"I think it's a dog."

"You sure?" Harolds squinted and pushed his glasses up the bridge of his nose, "Looks more like a drowned rat, if you ask me."

"He's too big for a rat. Go get some towels, Harold," said Irma adjusting her bathrobe,

"Towels? I don't think that such a good idea," Harold started for the laundry room. "We'll never get the smell out of them."

Harold continued to mumble out of earshot while Irma switched on the porch light to get a better look at their night visitor. The dog sat on his rump, shivering. His hair was long and black, which was unusual because he looked like an Irish Setter and Setter's were always red-haired.

"You know, I was just thinking," said Harold, returning with an armful of towels, "I'll bet I know whose dog that is."

"He's an unusual color, isn't he?" said Irma, thinking aloud.

"He sure is, and that's what got me thinking." Harold dropped the load of towels at his feet. "There ain't nobody who's got a dog that looks like that except for maybe Dave Murdock."

"I think you're right," Irma agreed. "He hasn't had the dog very long. Maybe a couple of months or so."

"Nope." Harold removed his glasses and attempted to clean them with the tail of his undershirt, but he only managed to make them worse by spreading the smudges around. "He's only had him since Gracie passed on, I think." Gracie had been Dave Murdock's wife. "I figure that dog can't be more than a couple of months old himself."

Irma pushed the screen open a crack and the dog slithered inside, burying himself in the towels. While Irma and Harold rubbed his coat, drying him off, the dog pounded his tail on the floor in gratitude.

"I swear, he's the only dog I've ever seen who could smile."

"He's probably hungry," Irma observed.

"We'll feed him and let him sleep, then I'll take him back tomorrow. "

The rain stopped sometime during the night leaving an abundance of nitrogen in its wake. This gave rise to a grassy perfume that lingered like a dense fog.

The next morning, Harold attached a leash to the dog's collar. They headed west, following the trail that banked the sloping shores of Hangman's Creek. The damp wind soughed through the valley, stirring life where it had lain dormant during the downpour.

When they reached a clearing, Harold stopped to adjust his gloves. He shivered against the damp morning air and took a deep breath, letting his head clear, slipping into a state of euphoric awareness. The air was cold but sweet. He watched the dog stand idly by, panting and gazing off into the distance.

The sky was clear and they could see for thirty miles in every direction. Overhead, a flock of mallards flew in tight formation, their wings beating in unison. The world was filled with the music of the wild; birds of different species tried to drown out the calls of one another. Softly, from distant farms they could hear cows lowing and horses whinny. Already, the air hummed with insects busy in their pursuits. The hills were wet with green and the earth breathed with life.

"I'll tell ya," Harold said to the dog, "if this isn't the prettiest goddamned place on earth, I'll put in with ya." Duke looked back at Harold as if he understood and Harold wondered for a moment if maybe he did. "And don't go telling Irma I took the Lord's name in vain either—you do and I'll be sleeping on the couch for a month." The dog yawned audibly and shook the dew from his coat. "Come on, we're not getting anywhere standing around here listening to you gab."

The dog seemed to know where he was going. He strained hard on the leash, pulling Harold behind him. His black coat shone in the dawning sun while his breath became vapor, appearing then vanishing, as he panted his way for home.

Dave Murdock's farm formed its own small valley—tucked between two knolls. Descending one of the hills, Harold noted the neglect the land had fallen into. The crops were unattended and the few animals Dave kept looked dangerously thin. The sorry state of the land became apparent as they drew nearer.

The dog tugged hard on the leash, taking Harold in tow and dragging him down the hill at a trot. Finally, tiring from the ordeal, Harold released the dog and watched him charge off for the farmhouse, barking, ears flapping, panting and dog grinning all the way. Harold caught himself smiling and shook his head; for some reason, that darn dog was a real kick in the pants.

As a child, their family never owned a dog; never owned a cat either. His sister had been allergic to most animals, so the closest Harold ever got to having a pet was raising a skunk.

He'd gone hunting with his father early one morning when the birds were just waking up and the air was so cold it rendered his nose insensitive to touch and made it run. His father told him to stop wiping his nose or . . . "it might break off and shatter like a soda cracker." Harold was only seven at the time and could never tell if his father was kidding or not; in this case he wasn't taking any chances.

It had been especially difficult for Harold to rise above the weight of overwhelming slumber in the autumn of 1917. The evening before, his older sister, Lillian, had read him to sleep. Mr. Tarkington's sensational new novel <u>Seventeen</u> was all the rage, and though the book had been out for nearly a year, Lillian had just received her copy from the library and was not about to put the book down, even when it came time to read her little brother to sleep. Lillian didn't seem to know when to stop reading. Harold didn't seem to know when to stop listening.

Lillian munched sunflower seeds and raisins while she read. With her knees drawn up to her chest, and her flannel nightgown pulled down to her toes, Lillian turned the pages with a snap, never looking up except when there was an especially titillating passage which she and Harold shared with delight. Harold had removed his spectacles after crawling into bed and lay smothered in his myopia—peering down a tunnel of unfocused space. He watched his sister's body sway to the rhythm of the words, her long chestnut pigtail trailing like a knotted rope. Behind her the candle cast her in an aura of haze and soft light. To his mind, Lillian looked like an angel washed in a shimmering halo.

"Papa said he might take us all to the flickers Friday next," she mentioned as she paused to snap to another page. "I can't remember the name of what's showing, but it's all the rage." Lillian resumed reading and Harold remembered the last time they all went to the Orpheum and sat facing a curtain while a man played a piano. Suddenly the lights dimmed, the curtain rose and moving pictures danced miraculously before his eyes on a silver screen.

Lillian's reading voice was smooth and rich with inflection. But, in spite of that, her voice soon trailed off into a nonsensical mumble of meaningless babble. Harold surrendered to the draw of sleep, sinking into his mattress. He didn't feel her lips on his forehead. "The sandman's on his way," she whispered.

In the morning he was nudged awake and within an hour he was on the trail with his father, hunting for small game. A cloud of damp air had snuggled down amongst the pines making the air wet. Harold pulled his cap down over his ears and shivered against the morning nip.

They happened upon the nest after Harold's father thought he spied a raccoon and fired his rifle. It hadn't been a raccoon after all, but a skunk. "Good Lord. I killed a polecat."

Her inert body lay next to an old rotting tree stump. At the stump's base, where the earth had been moved out, a tiny cave had been fashioned. Four baby skunks, about the size of Harold's small hand, huddled together for warmth.

"Shit," his father said, "now I've gone and done it."

Harold's father was carrying two rabbits, and a porcupine. He shifted the weight, and lowered his trophies to the ground.

They knelt over the nest and watched the little critters—their eyes still closed, sleeping, unaware. They squirmed for comfort when the wind whispered past.

Suddenly, Harold's father was fashioning his shirt into a sling.

"We're taking them with us?"

"We can't leave 'em here." His father had worn long underwear, so was comfortable enough once he'd fastened it at the neck. It was autumn, and nearly Thanksgiving—mighty chilly in the early morning hours.

"But, we can't have animals. What about Lillian?"

"Lillian needn't be bothered with them. We can keep them in the barn, and the rest of us can manage. If we leave them here some varmint will come along and eat them. We can't have that. It's not their fault. They'll need to be fed about every four hours or so, I expect. I'm not really sure what there is to feed them—maybe some warm milk with honey to begin with. I just don't know. I can't be there all the time, so I'm going to need your help, son."

"You can count on me," Harold assured his father.

That morning at breakfast, Harold's father announced his decision to the family. His mother and big sister Lillian were seated at the old oak table staring at the contents of the sling.

"You brought home four baby skunks?" Harold's mother asked aghast.

"Babies?" Lillian whispered, her eyes growing wet.

"Yes," Harold's father admitted, determined to own up to his responsibilities. "I admit it's my fault." He looked around the table, his face attempting to remain stoic, but his eyes growing wet. Inside, his heart ached. He believed deeply in the words his father had said to him when he was a child: *"God watches everything you do, son. Good and bad. Right now, in this life, is the only chance you'll have to correct anything you've done that was wrong—even if it was an accident and you didn't mean to do it. It's too late after you've died. If you're given the chance to fix it now, you better take it. It may be the only chance you get. God will love you for it."*

"Father says it's up to us to help the babies stay alive," said Harold, already having taken a great liking to the little polecats.

"Now, Lillian," Harold's father began again, "I know you can't be around animals, and I don't expect you to take part in their feeding necessarily."

"Oh, I want to help," Lillian interrupted. Lillian was wearing the same white flannel sleeping gown; her long braided hair hung over the back of the chair. "There's plenty I can do. I know a lot about polecats. I know their feeding habits . . . I can help prepare their meals; I can make sure whoever's turn it is to feed them, does it—there's a lot I can do."

"Thank you, Lillian," her father said.

"We'll all do our part," said Helen, Harold's mother. She caught her husband's eye and they smiled at one another. "Besides," she said, "the world's just too darn big a place to be alone in."

Fredrick Bridgeport, Harold's father, saw his wife as he always saw her— someone the Lord had taken great and wonderful care to create. He first looked into those dark green eyes in Palmer's Department store on Chicago's south side in the autumn of 1892.

She was standing behind a counter waiting on an elderly woman, and he had just stepped inside to get out of the rain. Fredrick didn't believe in love at first sight, refused to waste his time reading those silly romance novels that professed such things, and would have gone on believing and living the same lie had she not spoken to him when the elderly woman walked away. "Can I help you, sir?"

Rain pattered on the skylight and the air was chilled and damp. The muffled clatter of street traffic slipped under the doors and through the display windows. The hardwood floor creaked where he planted his weight.

Fredrick had just doffed his hat and stood running his fingers around the brim while staring into her eyes. He had never before seen eyes that green; wet, like emeralds in a brook. Her hair was heavy, a deep rich red and held off her shoulders with a barrette that matched her dress.

Fredrick was not a forward man. The few dates he had were prearranged, and because of his shyness nothing ever became of them. Above all, he was a gentleman and took great care to think before speaking—especially when speaking to a lady. "You have laughing eyes."

"Pardon me?" She studied him apprehensively at first, her forehead wrinkling slightly, then her eyes softened and her lips turned up as she attempted to suppress a smile.

Fredrick couldn't believe he'd just said that. He felt his face grow hot. He searched the distant shelves, his eyes darting about the store aimlessly, while his mind raced to think of something less stupid to say.

She watched his ears turn red—followed closely by the flush of the rest of his face. "Would you like to see what we have in the way of—umbrellas?"

"Yes!" he said a little too loudly. "Yes," he said again, regaining control, "that would be perfect. Just the thing when it's raining."

She smiled and brushed passed him. The scent of lilacs followed her and for a spell Fredrick thought he was losing his balance. What the hell's the matter with me, he thought. I don't act like this. Good God, she's going to think I'm some kind of a chump. He looked up and she was already several aisles away. He rushed to catch up—nearly running into her when she stopped suddenly, having arrived at the umbrella display.

She smiled and gestured with her hand, pointing them out.

"Yes," said Fredrick, "there they are." He plucked an umbrella from its stand and tested its balance. He held it up to the skylight and looked down its length as if it were a cue stick. "Straight enough," he announced. He really had no idea how to examine an umbrella—or if an umbrella was even inspected. "I love a good umbrella, don't you?"

"Especially when it's raining," she acknowledged.

"Yes, yes," Fredrick agreed, thrusting the umbrella forward with his arm outstretched, then recoiling. Suddenly he realized what he had done and flushed again. I must be some kind of an idiot, he scolded himself. What am I doing parrying this umbrella as though it were a sword?

"Touché," she said, smiling again.

My God, she has a beautiful mouth, he thought. "What? Oh, yes," Fredrick said, smiling back. I did look like I was fencing!—he admonished himself. I have to get out of here. Hopefully, before I make a total and utter fool of myself.

Yet, as desperately as Fredrick wished to be magically whisked away and be rid of the predicament he'd gotten himself into, he just as earnestly wished to remain— as near to her as possible. Before he could stop himself from speaking, he heard his voice say, "I'll take it."

As he walked back to the cash register, trailing, following in the smoothness of her walk, drunk from the scent of lilacs, charmed by her grace, caressing her brindled hair with his eyes, watching it hover between russet and sorrel, he felt himself sinking, drowning in an ache he couldn't name.

Fredrick had been attending night classes for two years at City College to become a Pharmacist. During the day he worked on campus, tutoring math to whoever could afford to pay, to make extra money. Not once had he felt alone—nor even homesick—until now. But, why now?

Walking back to the rooming-house, where he shared a flat with his friend, Ricardo Montoya, a pre-med student, Fredrick's sense of reality seemed dazed. He couldn't believe he'd actually bought the umbrella. Especially since it took almost every penny he owned.

An umbrella just wasn't his style. Fredrick was more inclined to wear a hat than to carry an umbrella. He was a hat man—it kept his hands free to carry books and what not. Though now, he had to admit, in this torrential downpour he was glad to have had the umbrella.

What the devil had gotten into him? Was he going daffy? Had his studies, classes and working finally driven him over the edge? Was the ever-increasing clamor of the city driving him bonkers? He put the question to his roommate.

17

Ricardo had listened to Fredrick's tale with stifled amusement. "It seems perfectly evident to me, old man," he said when Fredrick had finished.

"Well then, perhaps you won't mind taking a moment of your precious time to explain it to me, Ricky."

Ricardo overlooked his friend's irritation. He lit a cigarette and blew the smoke promptly out the window—smoking wasn't allowed in the boarding rooms. "You're love-struck."

"I'm...""Exactly."

"And, is that your diagnosis, dear doctor?" Fredrick asked, slumping carelessly into an over-stuffed chair.

"Well, you definitely have all the symptoms," Ricardo assured him. "Can you tell me a little more—I mean about her?"

"What's to tell? I don't even know her."

"Can you tell me her name?" Fredrick looked at Ricardo aghast. "Then I take it you don't even know what her name is." Fredrick shook his head, sighed and stared at the floor. "Can you describe her a little?" Fredrick rubbed his forehead, racking his brain for the words to describe her. "What color was her hair?" Ricardo prompted.

"Red."

"Well now, we"re finally getting somewhere! Eyes?"

"Green."

"Was she tall?"

"No," Fredrick said, hesitantly."

"No, " Ricardo repeated. "Short then."

"No," Fredrick said, hesitant again."

"Well then, could we say she was average?"

"Average!" Fredrick said, almost shouting and rising to his feet. "Certainly not. She was anything but average. She was exceptional, unique, first-class—fantastically beautiful." Fredrick began pacing the room. "Her hair had shades of red you've never seen before. She had eyes that could melt an iceberg, and the voice of an angel. She was exquisite in every detail and I don't think I've ever seen a more lovely woman before in my life. Nor shall I ever again." Fredrick slumped back into his chair out of breath. "She smelled good, too. Like lilacs."

Ricardo paused for a moment, then said: "Definitely love-struck." Ricardo checked his pocket-watch, snapping the cover closed. "But, there is a cure."

"Yeah, sure. And, what would that be—drowning?"

"The cure, my friend, is to take your mind off the matter. Completely."

"By doing what?"

"Well," said Ricardo, slipping on a coat. "I have to go out for about an hour to pick up some tickets. Do you like Beethoven?

"Yes, of course."

"Well then, listen to me. As yours pre-med doctor I prescribe an evening of exceptionally beautiful music to help ease your ills."

"I really can't afford a ticket right now, Ricky, " said Frederick eyeing the umbrella.

"I'll treat, and you pay me back when you can afford it." Fredrick thought a moment. "Music would be good.""Indeed. Music soothes the savage beast."

"All right, then. When and where?"

"Tonight, City Concert Hall," said Ricardo opening the door and donning his hat. "Oh, there is just one thing."

"There's a catch?"

"Just a small one. I promised Maria I'd take her to the concert as well." Maria was Ricardo's fiancé.

"Maria? Oh, of course. Is that all?"

"Yes," said Ricardo closing the door. "Oh," he said opening the door quickly again, "I also promised Maria that I'd find a date for her friend, Helen, for this evening too."

"What?"

"A blind-date is just what the doctor ordered, old man. I'll be back in plenty of time. Better freshen up," said Ricardo, slamming the door closed before Fredrick could summon a response.

The concert was scheduled to commence at 8:00 o'clock that evening. They assured one another they'd left in plenty of time, but the inclement weather had slowed traffic to a crawl. Though they were experiencing a break in the weather for the time being, nearly a million souls now inhabited Chicago. This density of humanity only served to make their journey more difficult—and seated now, as they were, on the horse-drawn trolley, it was plain Ricardo was growing anxious.

"Don't fret, Ricky," said Fredrick facetiously. "We're bound to get there eventually."

"I'm not worried about 'eventually,' I just want to get there before the damned concert begins." Ricardo ran his finger around the rim of his detachable starched collar. "Maria and her friend are not going to be happy."

"I'm sure they'll understand," Fredrick tried to assure him.

"Well, I'm sure it's easy for you to be so relaxed. After all, this is just a blind date to you."

"I prefer to think of it as medical therapy," said Fredrick, grinning.

Ricardo crossed his legs and shot his cuffs, ignoring Fredrick's wit for the moment. "Yes, but Maria is the woman I may ask to marry me one day—if she'll have me."

"Of course she will."

"Not if I continue to be late every time we plan a date," Ricardo said, exasperated. "I don't think I've ever been on time." Ricardo adjusted his hat for the umpteenth time. "At this rate, we'll never get there. Look here, the horsecar's stopped again! I think we could make better time on foot."

"I'm game if you are."

"Well, if you don't mind, old man, I say we give it a go."

In the middle of traffic, with horses, carriages, bicycles, pedestrians and a number of valve and piston gas-driven carriages, Fredrick and Ricardo jumped from the trolley and headed west, making their way up Garfield Boulevard.

Although it had been Ricardo's idea, Fredrick took to the task with a burst of energy—enthusiastically vaulting puddles, trotting along comfortably, alternating between the sidewalk and the street, seemingly unencumbered by the obstacles— invigorated by the exertion.

Fredrick was tall, with a long stride, and been on the varsity team since he was a sophomore. Ricardo, on the other hand, was shorter, had half the stride, only watched the varsity games from the bleachers and had a tough time keeping pace.

Two blocks from the concert hall it began to rain again. Ricardo cursed under his breath. Fredrick untied the strap to his new umbrella and hoisted it aloft. He felt himself smile.

It grew darker as dusk approached. Although the electric-light system was very near completion, flames of naphtha still blazed from nearly every storefront and some lamp-posts. A distinct hiss was plainly heard wherever rain seeped through the cracks in the lamp covers and dripped into the yellow flames.

"There they are," Ricardo whispered to Fredrick, a little out of breath. "By the sandwich-board marquee," he said, pointing.

Fredrick skimmed the crowd, his eyes skipping over unlikely prospects; finally coming to rest on two women. They were facing the street, looking off in the direction of the trolleys, their backs to the men, their faces a mystery. One wore a wide-brimmed hat and the other a poke bonnet with follow-me-lad ribbon streamers trailing behind it. Their hands were hidden in muff hand-warmers and their shoulders brushed one another for comfort against the wet and the wind as they chatted amiably. Their long pelerine capes nearly reached the sidewalk—one dark blue, the other dark green.

"I'm not sure this is such a good idea," said Fredrick, slowing his pace.

"Nonsense, this is just what the doctor ordered," Ricardo assured him.

"Yes, but . . ."

Ricardo stopped and looked at his friend. "You're not getting cold feet." "It's not that—it's hard to explain."

"Try."

"I guess—" Fredrick searched for the right words, trying to name the emotion. "I feel like I'm being unfaithful," he blurted. There, he'd said it.

Ricardo was speechless.

"That doesn't make any sense, does it?"

"Frankly, my friend—no."

21

"I didn't think so."

"Listen," said Ricardo, grasping Fredrick's arm, "try to forget this woman with the red hair. At least for this evening. Enjoy the concert. It'll all be over soon enough and a good night's sleep will see you right again."

"You're probably right."

"Of course I am." They resumed walking. "Try to be civil to this woman here, will you? I don't know anything about her, but I'm sure she's nice enough—Maria speaks highly of her. She works as a store clerk somewhere downtown. It's only for a couple of hours."

"Don't worry," said Fredrick. "I'll be the perfect gentleman."

As they drew nearer, Fredrick could feel his heart beat harder—not faster, just harder, as if it were about to burst through the walls of his chest. This always happened prior to meeting a strange woman—his anticipation, unfortunately being fed by his anxiety. Was she to be plain, beautiful, so-so? Would she be nervous too? Or, very sure of herself, making him all the more nervous? Would she be talkative—too talkative, and speak of boring things? Or, would she be silent, forcing him to think of topics to discuss?

"Maria!" Ricardo shouted, raising his hand.

They turned in unison.

Fredrick squeezed the handle of his umbrella until his knuckles were white. He never saw Maria; it was as though she didn't exist. Instead, his eyes were held captive by his blind date. Never did he imagine he would see those eyes again. Likewise, his blind date never noticed Ricardo; her eyes were fixed on Fredrick. When she saw the umbrella he was carrying, her face softened into a smile.

"Ricky," said Maria, taking Ricardo's arm. "I was afraid you were going to be late again."

"I'm sorry, but the traffic was perfectly dreadful. We got here as fast as we could. Perhaps you should introduce your friend," Ricardo suggested.

"Yes," said Maria, beaming. "Fredrick, this is Helen. Helen, this is Fredrick." She turned to Ricardo. "I guess that just about says it all, wouldn't you say?"

"Perfect," Ricardo agreed, smiling.

"Shall we?" She said, gesturing to the entrance of the concert hall.

They hurried up the steps as Ricardo searched his pockets for tickets. When the first four pockets turned up empty, he stopped and began rummaging through others.

"You didn't forget the tickets, did you?" Maria asked.

"Of course not," Ricardo insisted, checking another pocket. "A-ha!" he said, snapping his fingers. "Fredrick has them." He turned, expecting to find Fredrick and his date close by, but they weren't there. They'd shaken hands and hadn't let go.

"What the devil's going on there?"

"They don't know each other, do they?" Maria asked.

"No, of course not." Ricardo turned and watched them. Fredrick was unable to wipe the grin from his face as he spoke, and Helen suddenly threw her head back and laughed, grasping his sleeve as she did, to keep from falling. "At least, I don't think so," said Ricardo, musing.

At the conclusion of the concert, Ricardo suggested they all go at D'Journo's for a bite to eat before escorting their dates home. Fredrick and Helen lagged behind engaged in conversation, ever mindful of the rain that pattered the umbrella they shared.

"Do you think they enjoyed the show?" Ricardo asked Maria while vaulting a small puddle.

"Who could tell? They whispered to one another throughout."

"Yes," Ricardo agreed, "they did seem to visit a lot."

"A lot? Why they never stopped. I doubt they heard a single note Mr. Beethoven wrote?"

They had—most of it anyway, and each note was all the more precious because of their rendezvous. And sometime during the concert he took Helen's hand and held it. When they married, they both agreed Beethoven's "Moonlight Sonata" should be played at their reception—and it was.

Nineteen years later, seated across from him at the old oak table, Helen's eyes were looking back at him—just as moving, just as beautiful as always. She was flanked on either side by their children.

In spite of her allergies to animal hair, Lillian felt compelled to touch the critters and her sensitivity to animal hair, in time, diminished to a tolerability that satisfied both her comfort and her maternal sensibilities.

Though they all left when they were grown to roam the wilds of the woods, sooner or later they each returned for meals or affection. Every spring or fall the females reappeared to show off their new litters; their broods learning at the same time where to go for help or a good meal if need be.

Those were the only pets Harold knew growing up. And though the experience was satisfying, it was fleeting. The skunks were only partly domesticated, still drawn to the wild. What he had always wanted was a dog. His own dog. Thirty-seven years later, it's what he still wanted.

Now, as he descended the miry incline, he couldn't help but smile as he watched Dave Murdock's dog—the leash still attached and trailing behind him, hightailing it for the farmhouse, his feet kicking up dust and mud as he headed for home.

When Harold reached the floor of the valley, he kicked the mud from his boots. So much had collected by that time, it felt like he was lugging an extra five or six pounds on each foot. It made his back ache and shot sciatic throbs down his leg. If he still smoked he would have lit a cigarette now just to dull the pain, but he'd quit years before. Instead he unwrapped a stick of wild cherry chewing gum, rolled it into a ball and popped it into his mouth.

As he approached the farmhouse, he saw how thin Dave's two horses were. The ground was mostly bare where they'd exhausted the land of grass, and their water-trough was dry. It wasn't like Dave to neglect his animals. He didn't own many—a couple of horses, a milk cow, a few pigs, some wild chickens, and of course, his dog. But what animals he had, he loved and cared for.

Harold decided to stop and feed the horses. It only took a few minutes and he knew he would hate himself later if he didn't take the time to do it now. Maybe Dave is ill, he thought. He filled their water-trough from the nearby pump then stood and watched them quench their thirsts.

He was turning away, starting for the farmhouse again, when he was met by Dave's dog. He came trotting up with his tail between his legs and his ears laid back. If Harold hadn't known better he would've sworn he read concern in the dog's eyes. Something was wrong—very wrong.

Together they made their way to the farmhouse. The dog kept close to Harold—his body pressed against Harold's leg as he trotted alongside. Harold stopped when they reached the back of the house and peered in the window. Dave was seated at the dining table, his back to Harold. He rapped on the glass. "Hey Dave! It's Harold Bridgeport from up the road!" Dave didn't budge.

The feminine charm that once graced the Murdock homestead was gone. The doilies, the place mats, the fresh flowers that always occupied one or more vases, the tidy arrangement of bric-a-brac—Gracie's collection of Chinamen—porcelain, wood, and brass—all gone now.

Harold and Dave had been neighbors for seven years, plenty of time to get to know and appreciate one another. When Harold had heard new neighbors were moving into the farm down the road, he decided to be a good neighbor and introduce himself.

Thinking back on it now, he realized it had been Irma's idea to introduce themselves, and it was she who had gone to the trouble to make one of her famous coconut cakes as a housewarming gift—white cake with white frosting and liberally sprinkled with coconut.

Irma's cakes were taller than normal. Her secret was to make enough batter to fill three cake pans instead of two. She made her frosting especially creamy by blending melted marshmallow and cream cheese, then made giant peaks and waves when she iced. Though her new neighbors were unaware of the honor, both Gracie and Dave could tell there was something special about that cake. Fact was, Irma's cake had won a blue ribbon at the Spokane County Fair for two years running.

They sat the shade of the font porch on unpacked cardboard boxes. It was mid-afternoon on a Sunday and the Murdock's new home was awash in shadows. Their three sons, Charlie, Jed and Josh, five, twelve, and. fourteen were rough-housing up by the barn—their banter unintelligible because of the distance.

"Come on up here, Charlie," Jed coaxed.

"No thank you, brother dear," Charlie answered. "You're not a coward, are ya?"

Charlie turned to Josh who was standing next to his little brother. Jed had always been the crazy one, always the first to take a chance— willing to accept a dare no matter how foolhardy.

"Well?" Jed yelled down. "Are you a coward or aren't ya?"

Jed had managed to pull himself onto the upper hayloft using the rope that was attached to a pulley on the outside of the barn. He'd threaded one end of the rope through the belt loops of this overalls then hauled himself aloft, pulling on the other end. "Am I a genius, or what?" he exclaimed as his body rose higher off the ground. "I do believe I'm one hell of a damn smart cookie, you know that?" The exertion had forced him to grunt, which embarrassed him. "I call this 'The Big Pull'," he announced. His arms trembled, nearly giving up on him, while tiny beads of sweat oozed from his young pores.

Now that he had made it—now that he had successfully demonstrated his skill utilizing the 'Big Pull' and managed to scale the side of the barn, Jed insisted his brothers attempt the same ordeal. "Come on, you cowards. Climb on up here."

"You know, Jed," said Josh, using his hand to shade his eyes from the sun, "you're about the dumbest genius I know."

"What's that supposed to mean?" Jed yelled.

"Why should we pull ourselves up with a rope?" Josh asked grinning.

"How else do you think you're ever gonna get up here, big brother."

"Oh, I don't know," said Josh. "I thought Charlie and I might climb the ladder instead."

"What ladder?"

"The one just inside. Come on, Charlie." Josh guided his little brother into the barn and they started up the ladder. They could hear Jed still shouting at them, his voice muffled by the dense barn walls.

"What ladder?" Jed bellowed. "Anybody can climb a ladder. Who wants to climb a stupid ol' ladder, anyway?"

"We're not cowards, are we?" Charlie asked, stopping for a moment to peer down at his brother, Josh.

"No, we're not cowards, Charlie," Josh assured him.

In the shade of the front porch, Harold sat next to Irma and watched the two boys disappear into the barn. They soon reappeared in an opening near the top of the barn by the upper hayloft and joined their brother who was still tied to his makeshift elevator.

Gracie had served up slices of cake on the paper plates Irma had brought along. Harold had made a pot of coffee and poured them each a paper cup from a thermos. They sat and ate and chatted and watched the barnyard slip deeper into shadows as the sun moved farther west.

"Sure is pretty here," Dave Murdock remarked.

"Prettiest God—" Harold caught himself and felt Irma's eyes on him, "—blessed place on earth."

That had been seven years ago.

Harold rapped on the window again. "Dave?" Still no response. Harold shrugged and rounded the house to the front door. He knew Dave was hard of hearing.

The front door was unlocked, so Harold let himself in. "Dave?" The dog stayed close to Harold, panting, still hugging his leg.

There were no lights on, but plenty of daylight poured in through the curtainless windows. Dave had taken the curtains down when his wife died. He couldn't stand looking at them. It was too painful. He remembered when Gracie made them, and he knew they looked really nice hanging on the windows—but since she'd passed away, he didn't care if anything ever looked nice again.

All of his sons were gone now. Jed had joined the Marines to fight the Japs in the Pacific; Josh had joined the Infantry to liberate Europe. They wrote home when they had the urge. Josh wrote more letters than Jed. Dave knew that for a fact because he kept his sons' letters in separate shoe boxes, rubber banded together in neat stacks. When he finished reading one, he'd reread it, then count how many letters he had accumulated—for Dave, it was just as good as counting money, maybe better.

And then there was Charlie. That's what killed Gracie—Charlie. The baby of the family became the most precious—the one all of the hopes and dreams for the future were invested in. They expected great things out of Charlie. He was going to be a lawyer or a doctor or the statesman of the family. He was handsome enough to be a movie star. He was bright, too; his intellect shined.

But, Charlie had one flaw. Because he was svelte and naturally gentle he feared he was inadequate. The incessant teasing by his older brother, Jed, only served to make matters worse. Soon, Charlie came to doubt his manhood and felt compelled to prove himself worthy by doing whatever it is young men think worthiness requires of them.

One day, while Gracie was making lunch, standing at the counter by the enamel-chipped sink, slicing tomatoes to garnish the sandwich plates, she looked out the window and watched Charlie climbing to the roof of the barn. What he was trying to prove by attempting such a foolhardy quest remains a mystery.

She remembered pausing, concerned for his safety. It had rained the night before and the shingles were still wet and the moss that grew there was slicker than snot. Charlie straightened, and stood tall, planting his hands on his hips in triumph. He was so proud of himself.

Then she watched in horror as Charlie lost his footing and slipped, sliding down the slanted roof, his fingers desperately grasping for anything to stop the momentum of his descent.

He looked around, his eyes jerking from apex to edge, frantically scanning the horizon. He could see everything from up there—the wooden tracts of pines that stretched for scores of miles through the valley, even the many creeks and streams that wound without reason like confused snakes through the dips and gullies.

Everything's going to be alright, Charlie told himself as he toppled over the edge and began to free fall. There's no need to panic, no reason to fear—Charlie was sure he was going to be all right. He was falling so fast he felt like he was flying—diving like a pigeon-hawk. He imagined himself flexing his tail feathers at the last moment and swooping skyward to escape injury, averting his peril.

He was nearly upside down when he twisted his head to catch a glimpse of the farmhouse and the steep knoll in the back where a trail wound past Hangman's Creek.

"Come on, Charlie, you little fraidy-cat!" Jed had waded out into the middle of the creek and stood atop a large boulder. "I'm gonna teach you how to fish, but first you're gonna have to get your candy-ass into the water."

Charlie rolled up his pants, slipped off his shoes and cautiously stepped into the rushing water. It was cold and he withdrew his foot.

"Come on, you pansy!" Jed shouted. "You're never gonna catch any fish stalling." Jed snapped his fishing pole and cast his line. The lure jettisoned across the surface of the choppy stream.

"I'm not stalling," Charlie insisted and once again he stepped into the moving water. The rocks were slick, but Charlie persisted, immersing both feet now.

Hangman's Creek had varying moods depending on the time of the year.

Sometimes she was calm and the water flowed over her rocks like gentle words; sometimes her water was low and her banks dry from the summer's heat. But, in the spring she was pregnant with high water from the mountains and gave birth to small rapids and eddies that wrapped around Charlie's leg and pulled him in.

Charlie scrambled to catch hold of the tall grass that grew along the banks. "I'm sorry, Charlie, but I'm afraid I just can't teach a clumsy oaf like you how to fish. It just wouldn't be proper," Jed rallied, turning away and laughing.

Charlie's finger-tips brushed the blades of a clump of grass as he frantically stretched, reaching to grasp hold . . . much like his fingers attempted to break his fall and brushed by the rain-gutter as he slid past. The last thing he heard was his belt buckle clicking, stumbling over the slats, and the air rushing past his ears as he left the roof of the barn.

Gracie watched Charlie's body twist as it fell silently to the ground and she stopped slicing her tomato, frozen. Gracie lived that moment over and over for about a year. She gave up looking for reasons to live after that. She didn't bathe as often, nor eat as well. She grew thin and became susceptible to every vagabond bacterium or viral mutation that dwelled within the Murdock estate—finally succumbing to meningitis which attacked her without prelude and killed her without mercy.

Maybe they all would have been better off if Charlie had simply died. Gracie might have been able to handle that—that's what Dave Murdock thought. But as it was, Charlie fractured several vertebrae in his neck, broke an arm and a leg and several ribs too, but that was nothing compared to the damage to his brain. They couldn't repair that, but they knew how to keep him alive—and keep him alive they did.

They put Charlie is an old folk's home, because it was the only way the state could afford to keep him. It was a cinch Dave couldn't afford to pay for Charlie's care. They stuck tubes up his nose so they could drip food into his belly. Eventually, they just stuck a tube directly into his stomach and wrapped diapers around his butt so he wouldn't defecate all over the sheets. Charlie became a vegetable. He didn't know up from down. In Charlie's world he was still falling, even flying occasionally and admiring the scenery—remembering the 'good ol' days.'

All Dave had left after Gracie passed away was his dog and a few farm animals to keep him company.

"Dave?" Harold pushed the door open and stepped over the threshold—next to him, the dog stopped panting and walked in too, his nails clicking on the wooden floor. That was when the odor hit him—an overwhelming stench so thick and rank that it made the air wet with its presence. It was like sticking his head in a barrel of meat gone bad. Harold's stomach turned and he found himself back outside, retching into the dirt.

When he recovered, he got slowly to his feet, dazed. Blow-flies buzzed passed his head. They were coming from inside. One landed on his neck and Harold slapped it away in disgust. Dave's dog was getting nervous. He pawed the ground and barked, then whimpered a little. Harold reached down and patted his side and assured him that everything was going to be all right.

But, everything was not all right. The mephitic stench, the untended animals, the blow-flies, they all made sense now. Harold did't have to go back inside to ally his suspicions, only to confirm them.

He gathered his courage, took a deep breath of the cold morning air, then tapped the door open with his boot. The dog stayed behind, refusing to enter, whimpering now and then.

Dave was seated at the dining table, just as Harold had observed him earlier from the window. He'd been dead for some time. Long enough for blow-flies to lay their eggs, hatch and mature. Several generations were feeding on Dave's carcass. A few blind larvae wriggled helplessly on the table.

A wet breeze drifted through the house, lifting wisps of Dave's hair and flipping the edges of a telegram on the table. A coffee mug held it in place, preventing it from riding the air currents and sailing across the room. Harold watched it flip and flap—beckoning in perfect meter, a gentle cadence of waves. Harold waded into the room.

The coffee mug was easily within Dave's reach—but Dave wasn't reaching anymore. Harold tip-toed to the table, slid the telegram free and hurried back outside—relieved that it was over, grateful for the fresh air.

It was from the U.S. Government. They regretted to inform Dave Murdock of the death of his son Joshua. He'd been killed at Normandy during the allied invasion. What the message didn't say was Josh never quite made to shore. His body crumpled over and fell into the cold waters of the Channel the instant he left his landing vehicle. Now Jed was the only one left—and that wasn't enough to keep Dave's heart from finally breaking.

The Sheriff arrived at Irma and Harold's a couple of days later. Irma was in the kitchen whipping cream. Two bowls, already brimming with fresh strawberries sat nearby. Irma took down another bowel when she saw the Sheriff's car drive up.

Harold was sitting at the kitchen table reading a farming journal and enjoying a cup of coffee. Dave's dog was under Harold's chair, sleeping.

The Sheriff had stopped by to give them an update. After all, Harold had found Dave Murdock's body and reported it.

"Coroner says it was probably a heart attack—near's he can figure," said the Sheriff accepting a bowl of strawberries smothered in whipped cream. "Thank you, Irma." The scent of vanilla hung in the kitchen air.

The Sheriff was a big man, played line-backer in high school and college, and had put on a number of pounds since then. His butt could have filled two of Irma's kitchen chairs. "Talked to the judge too, He says since nobody can be reached—you know, what with Charlie being in the condition he's in, and Jed being somewhere off in the Pacific—that the county has to probate the estate."

"What exactly does that mean?" Harold asked, watching Sheriff Kendale eat. He couldn't remember seeing a man put as much food in his mouth at one time as the Sheriff did.

"Well," he said, "everyone knows you've been tending and feeding Dave's animals since his body was discovered. The judge thinks you should just take charge of them. The county will assign custody of the animals over to you, if you want 'em."

Harold considered that for a moment. "What about when Jed comes home after the war."

"You mean if he ever makes it out alive." The Sheriff inhaled another mouthful of strawberries. The traces of whipped cream that remained on his lips were quickly wiped away with the back of his hand. "Apparently, according the judge anyway," he continued, "the fact that you're taking care of them will warrant ownership.

"If Jed decides he wants them back, he'll have to pay a boarding fee for every day you've kept those animals alive. Most likely that'd be a heck of a lot more than they're worth. Besides, everybody knows Jed doesn't give a shit about those animals. Sorry, Irma."

"That's all right. I've heard worse language than that around here," said Irma, eyeing Harold.

"So," said Harold, looking under the chair, "I guess that means I've got myself a dog, then."

"Looks that way," he said, eyeing his empty bowl regretfully. Harold wondered if the sheriff was going to pick it up and lick it clean. "What are you gonna name him?" He hesitated, then shrugged what the hell, picked up his bowl and waved it at Irma. "You got enough for another?"

Irma just nodded and filled another bowl with strawberries.

"I don't know," said Harold musing and watching Irma dish oodles of whipped cream over the strawberries. "I like the name 'Duke'. What do you think, Duke?"

Duke pounded the floor with his tail and looked up at Harold, smiling. Now, Duke was just another part of the family.

That was eight years ago.

Jeffery, who was staying in a little bedroom above Irma and Harold's back porch, rolled over and pulled the covers up tight to ward off the night's chill.

Jeffery heard his aunt and uncle leave the porch and go into the house. Maybe the quiet would let him sleep, now. But, it didn't help much. Eventually he drifted off, but even while he slept he thought he was awake.

The next morning, before breakfast, Harold and Duke were in the new chicken coop, setting up shop for another day of building. Duke pranced around with a hammer in his mouth thinking he was helping. Harold was moving a saw horse into a better position when the sound of the screen door slamming made him look up.

Breakfast was about an hour away, and though he wasn't obliged to perform any morning chores, he usually kept Harold company while he set up shop.

But, today was different. Today was the day of the family picnic and before his relatives began arriving—pulling up the drive in their big family cars, the tires spinning and kicking up gravel, he needed time to think.

If it hadn't been for Nellie's surprise visit the day before, he wouldn't have felt compelled to ponder now. Ruminating leads to reminiscing—dredging up the past; a past Jeffery took pains to forget. Growing up was difficult enough without entertaining the images of his tortured childhood. But, Nellie had said something he couldn't stop thinking about.

The sun rose over the alfalfa fields, basking the barn in soft sunlight. The warmth, coming through the slats, smoothed the goose bumps from Jeffery's legs as he closed his eyes and remembered the day before.

Nellie Fairfax, Jeffery's cousin, came from the dairy farm down the road. Her parents owned the only dairy farm within fifty miles of Colfax. Short, teasing, little Nellie always stood her ground with her legs planted firmly apart and her hands on her hips. Nellie, with her pug nose and the smart-ass whiny voice just couldn't leave her visiting cousin alone.

She was wearing a thin polka dot dress that rose just above her dirty pink knees. That's all she was wearing, too. She was even barefoot.

All Jeffery wanted to do was bounce on Uncle Harold's good-smelling leather seat—bounce and pretend like he was steering that big old rusty tractor that hadn't seen a lick of work in forty years. Just smell the hay in the barn and watch the swallows dart in and out of the giant open doors—swooping into the shade of the cool interior from the blistering heat of the midday sun. Bounce, while hanging onto the giant black steering wheel—sailing through the fields of ripe wheat or corn or barley as the large treaded tires rolled over the warm earth, stirring up mounds of soil—pretending like he was plowing a field, while . . .

"Hey, shit-head! What'cha doin'? Playin' farmer?"

Jeffery jerked upright, embarrassed. "Who let you in here?" An enormous wave of heat washed over him. In that moment he could have killed cute little Nellie Fairfax. He had the urge to jump off that old tractor, grab a pitchfork, and spear her right through the belly—right through her pretty polka dot dress.

Nellie turned her back to him and bent over, talking to him from between her legs. "I'm goin' down to the creek to swing."

Jeffery sat for a moment, stunned. He stared at her bare bottom then quickly looked away, examining the over-sized steering wheel. Nellie was crazy like that.

She turned and straightening up. "You comin'?"

"Yeah," he finally said, jumping off the tractor. "I guess so." He attempted to sound bored, reluctant to go along with Nellie—who needed her anyway?

He was damned if he was going to let her know how desperately he needed the company, or how anxious he was to grab hold of that thick knotted rope. Uncle Harold had smuggled it off an Ocean Liner some twenty-odd years earlier. He tied it to the beetling limb of an old tree that grew along the bank of Hangman's Creek.

The rope was so thick Jeffery could barely wrap his hands around it. Even so, it left invisible slivers of hemp embedded in his palms and his shoulders would ache from supporting his weight as he swung out over Hangman's Creek with its swift moving water and fine polished stones.

The creek was shaded on either side by twisting oak trees that had been growing there since anyone could remember. Along the path of its winding watercourse grew a thicket of brush, wild flowers and weeds that danced to the constant breezes drawn to the creek's banks. It was chilly by the water's edge.

Jeffery pushed his hands into his pockets and watched his sneakers shuffle along the beaten path while Nellie skipped beside him, puerile, and talkative, swinging her arms. He tried ignoring her, intimidated by her vivacity, but decided she wasn't really all that bad. For a cousin she was all right—a little overbearing at times, forever getting into mischief and dragging whoever had tagged along into the same predicament. But, like most of the cousins, and there were fourteen to contend with, she was loyal and would fight tooth and nail for your honor if it came to that.

"How long you here for?"

"All summer, I guess."

"All summer?!" Nellie slapped her forehead. "Jesus, you mean I got to put up with your ignorance all summer?" Jeffery scoffed and Nellie patted him on the shoulder and laughed. "Don't take everything so seriously, shit-head."

"Quit calling me that."

"My, aren't we sensitive today." Nellie kept her hand on Jeffery's shoulder. "Okay, how 'bout if I call you booger-nose, instead?" Jeffery raised his fist and feigned striking.

"All right, butt-face then, but that's my final offer." Jeffery turned and started back to the barn.

"All right, all right. I take it all back." Jeffery stopped. "I won't call you any of those things." He turned, easily appeased, and joined Nellie again. They walked on in silence. Then—"would pig-fucker offend you?"

Jeffery couldn't contain himself any longer and burst into laughter. Nellie, too, thought herself quite amusing and hung onto Jeffery to keep from falling down as a wave of merriment gushed through her.

When they arrived at Hangman's Creek it was awash in shadows. Sunlight snaked through the leafy mantel when the leaves moved. The water was swift and cold. On hot summer days, the cousins would have contests to see who could stand barefoot in the frigid water the longest.

Harold told them the water that ended up in Hangman's Creek came from the ice melting off the Cascades, hundreds of miles to the west. By the time it reached Spokane it was cold, clear and clean from rushing over countless rocks and pebbles.

Nellie immediately mounted the rope and swung out over the water.

Jeffery squatted on the bank and poked a stick under a submerged rock; he grinned as the disturbed tadpoles bolted for safety.

"So, will you be here for the family picnic?"

Jeffery shrugged. "I guess so. As far as I know, anyway. Who's gonna be here?"

Nellie was straddling the rope. She pumped her legs and swung dreamily over Hangman's Creek. The danger of being suspended high above the rocks and rushing water and the thrill of G-forces pushing against her as she swung through space, brought contentment to her spirit. "Who's coming?" she repeated, absently. "Everybody, I guess. After all, it is a family picnic."

As the rope sailed lazily the forced air blew through Nellie's red hair and lifted her dress. She extended her arms and leaned back, peering up through the leaves at the sky as woolpacks drifted past. "I'll bet everybody ain't gonna be there," she heard Jeffery say. She recognized a note of sadness in his voice. No, she thought, you're right, Jeffery, probably not everybody.

Looking down, she could see the top of Jeffery's head, his platinum hair neatly trimmed and parted, but a little too long where it fell across his forehead. His skin was so white it was almost ivory, and his eyes so deeply turquoise it was hard to believe they were real. When she allowed herself to admit how drawn she was to Jeffery, something stirred in her stomach like butterflies. Nellie shook her head and blushed. It's unnatural to feel this way about a cousin, she thought, I'm probably mentally disturbed.

"My mother's not going to be there, is she?"

"How should I know?"'

"I haven't seen her in five years."

Nellie twisted her body and made the rope spin, hoping she'd get so dizzy she'd almost loose her balance. The trick was not to get too dizzy, then you really could lose your balance. Cousin Paul did that last summer and nearly broke his neck.

It was hotter than a firecracker on that day. Paul was sweating like a horse by the time they reached the small clearing near Hangman's Creek. They could already hear the water rushing over the rocks. In another month the water's depth would recede —but for now, it was high and swift.

Six of the cousins journeyed to the creek on that occasion. Nellie and Theresa flanked Jeffery. Philip, Paul and Steve tagged along at the last minute.

It was as though they all became aware of the rope dangling over the creek in the same instant—as if there was a psychic bond. No one had to say anything; it was understood that the race was on.

Paul was the first to break free from the gang. He was tall for his age with long red hair so bright it was almost orange. His hands were large, the long delicate fingers made ugly by nails he had bitten to the quick. He needed glasses, but refused to wear them and walked with a gait that suggested a disdainfully proud mien. This facade he adopted as a line of defense. His feelings were easily hurt, though he never showed it. Instead, he recited standing phrases to convince others of his toughness or apathy: "Hey, I could care less." or "So, who really gives a shit?"

They heard Paul yell over his shoulder as he broke into a run, "last one there has to kiss my white ass!"

There was a last-second scramble to reach the rope first. Steve, Philip and Jeffery nearly knocked one another over as they struggled to get their feet moving. In the melee, Theresa was pushed aside and fell. Nellie stopped and helped her up while the guys rushed ahead to the sloping banks.

"I hope they break their necks," said Theresa getting to her feet and brushing the dirt from her legs. She frowned at the injury to her knees.

"Don't worry," Nellie assured her. "With any luck, they probably will."

Paul reached the rope first. He dove recklessly in a desperate leap and caught hold, his toes skipping across the surface on the crest of a ripple. His fingers wrapped tight around the woven hemp as he swung out over the rushing water, wearing a Cheshire grin. "Hey, look at me!" Paul shouted. "The Winner!"

Using the strength in his arms, Paul flipped his body. Now his feet pointed to the sky and his head faced the moving water. "I'm the winner!"

"He means he's the wiener," said Philip.

Nellie chuckled. Steve and Philip were the comics of the troupe.

"I'm pooped," said Theresa sinking to the ground.

"Well, there ya go," said Philip. "Your sister's pooped and Paul's a wiener. I'm not sure our parents would approve of the company we keep."

"I'm not sure Paul has a wiener," said Steve.

"I'll ask him," Theresa offered.

"Oh, he's got one alright," said Nellie. "I've seen it."

"Well, why don't you describe to for us, Nellie," Steve prompted

"Hey, Paul," Philip hollered. "Have you been showing my sister your wiener again?"

"What?" Straddling the rope now, Paul could barely make out what the others were saying. Most of it was muffled by the roar of the water. He twisted his body upright and threw his weight in an arc, allowing gravity and momentum to wind the rope.

"Has anybody got a light?" Jeffery asked, pulling a cigarette from his shirt pocket.

"Jeffery! Where did get that?" Nellie gasped with delight.

They gathered around and Steve whipped out a Zippo.

"Aren't we gonna wait for Paul?" Theresa asked.

"Paul ain't a smoker," Philip informed them. The wick ignited and the cigarette was lit.

"That's right," Steve confirmed, "he'd rather be a wiener."

He accepted the cigarette, took a puff and passed it on.

"Ain't that right, Paul!" he yelled.

"What?" His cousin's voices were drowned out by the wind whistling past his ears. The rope had begun to unwind, gaining momentum, each spin faster than the last. The noise from the river became a roar and the centrifugal force pulled his cheeks away from his mouth. The world whipped by in a blur of leaves, trees and hot summer weeds.

Jeffery watched from the bank as Paul became a bleary, spinning top. "He'd better be careful," said Jeffery, accepting his turn at the smoldering weed and nodding to the creek. The others turned to watch.

Paul was wearing a red T-shirt and green shorts. Spinning like he was, these colors blurred into soft lines—just like the top Jeffery sometimes played with on the kitchen floor when he was younger. When it reached its optimum spinning rate, the top would hum and the pictures painted around the perimeter dissolved into a soft rainbow of orange—red—and green.

There, flanked by his cousins they watched cousin Paul spin like a top. Jeffery marveled at the acceleration Paul had somehow managed to attain. It seemed impossible that anybody could muster the strength to cling to that massive rope-swing as it continued to unwind. His ever-increasing velocity produced its own haunting sound—a hum that hung in the air, audible even over the roar of the rapids.

"Damn!" Steve said in a hoarse whisper. "Look at him go."

"Ride 'em cowboy!" Philip shouted.

Jeffery blew out his smoke, and crushed the butt into the dirt. "I hope he's strong enough to hold on. I'm not sure I could."

It was then that Nellie realized there wasn't going to be a happy ending to this story. Jeffery was definitely the strongest of the lot. If he doubted his ability to hang onto the rope, there was no way Paul would be equal to the task.

Paul's body slowly began to pull away. Try as he might, his arms couldn't compete with the forces of nature. Every muscle strained, his fingers felt like they were being torn from his hands and his knuckles burned as he tried desperately to draw the rope closer—but it was no use. Paul was soon yanked away from the whirling rope and hurled into the air like a discus.

Paul's body was still airborne when his cousins reacted. Jeff was the first to leap from the bank's edge and plow into the creek. Nellie and Steve followed close behind. Philip and Theresa rose to their feet but remained frozen and wide-eyed.

Paul landed in the water on his back and sank like a rock. When he opened his eyes he could see the water rushing over his head and the rippling image of his cousins peering down at him. He panicked when the water swept up his nose and down his throat.

Yep, Paul was damned lucky his cousins had been there to pull him out. By the time they reached him, he was already sucking water up his nose and bleeding like a stuck pig. Jeffery whipped off his T-shirt and held it to Paul's head where it was split open. They helped him back to the farm where Irma ordered Harold to drive Paul into Spokane to emergency. There he received fifteen stitches.

A year later, sailing lazily over Hangman's Creek, Nellie chuckled quietly, remembering that day. Below, Jeffery watched her while lying on his back, his hands folded behind his head.

"Are you enjoying the view, Jeff?"

"Yes, ma'am, I surely am." Jeffery smiled and then added, "you're pretty much covered up, now."

Nellie pulled with her arms and stood with her feet on the knot. The polka-dot dress billowed, and Nellie showed herself. "Is that better?"

"Much better, thank you."

"You're welcome, sir." She feigned a curtsy.

"So, tomorrow's the family picnic, huh?" Jeffery asked.

"Yep. You gonna be here?"

"Beats me. I never know from one day to the next where I'll be. It depends."

"On whaat?" Nellie leaned back and squinted at the sunlight filtering through the green overhead.

"Well," Jeffery began, sitting up. "the way I got it figured, if my mother's been invited, they'll shuttle me off to stay with a sick aunt or something for the day." He poked his stick into a pile of fallen leaves, black beetles, earwigs and roly-polies scuttled off into the shadows. "And, if she ain't been invited, I'll probably be here for the picnic."

Nellie leaned to one side and encouraged the rope to begin winding up.

"It's been nearly five years," Jeffery muttered.

The rope unwound a little faster than Nellie had anticipated and she nearly lost her grip. "Crap." Nellie righted herself and waited for her heart to slow down. "It's been five years for what?"

Jeffery looked up at Nellie incredulously. "Since I've seen my mother." Jeffery lay back again and locked his fingers behind his head.

Nellie pumped her legs to increase the arc of her swing. "I guess five years is an awful long time," she admitted. "Are you sure you remember what she looks like?"

"You never forget what your mother looks like." Even if you were only four when you saw her last, Jeffery thought.

When you're four, a day can last a lifetime. A memory, whether real or imagined, becomes truth. Looking back is like sinking into a cushioned seat in an empty theater and waiting for the curtain to rise.

In Jeffery's movie-memory the curtain rises without credits and finds him on a wintry country road at first light. The wind brushes his hair as the cold numbs his nose, and music, barely audible, rises painfully out of the depths of white silence. Nature whispers to him in the voices of soft morning birds and a brisk winter breeze.

He's a small boy, mittens safety-pinned to his wool coat, standing quietly, patiently. His wool cap, forced tightly over his head earlier, now rests on the crown of his head—cocked back, the way he likes it—the wool, warm and itchy. His hair, a little too long, but fine and platinum, falls over his forehead and he refuses to brush it away from his eyes.

A light wind lifts powdery snow that drifts up his back and glides smoothly over his head. It falls slowly—floating down before his knitted brows and parakeet-blue eyes. He looks up the road and listens with protrusive snow-white ears for the sound of a car that never comes.

"Jeff!"

He pretends not to hear, ignoring the voice of his aunt. Snow begins to fall again. Just a few flakes, large ones—with the promise of more to come.

Jeffery looks up the road again. Nothing. Just the knob of a hill where the fence disappears over the ridge near a bare tree all covered with a snow so fine it forms small clouds whenever the wind blows and hovers in the air just above the branches like a flock of miniature seagulls.

But, there are no seagulls—not here in Newport, Washington. *"We're too far from the sea and too close to the sun."* That's what Grandpa would say.

Last summer there'd been seagulls—so many it would have taken a four-year old all day to count them. Savannah, Georgia had grasshoppers so big they could jump over rooftops and rains so warm they dropped tadpoles on your head when you weren't looking. Savannah had white houses with no basements and a mother with deep auburn hair who stood on the front porch in a summer dress drinking iced coffee, the cubes clinking like wind chimes whenever she sipped.

The snow had drifted into mounds and he sunk up to his knees as he turned and trudged back to the farmhouse.

Breakfast was a blur. His eyes wandered to his overcoat, his armor against the harshness of the cold and the wet. It hung on a nail by the backdoor, the snow melting and making a puddle where it dripped.

Aunt Jackie was standing at the stove scooping up the sticky mess she called oatmeal. "Sit yourself down before it gets cold." Jeffery was already seated. Her hips swayed back and forth as she worked and Jeffery was sure he could see her rump through her thin cotton dress.

His father limped clumsily into the kitchen favoring his right leg, his work boots heavy on the wooden floor. His face was thin and solemn, giving the impression he never once smiled in his life. He sat down hard and scooted up to the table. Jeffery watched his father's eyes avoid his own, and waited for the insects to die.

"Feelings are like little bugs, boy. They crawl around in you, just as content as a calf in the hay. They live their own little lives and some can live forever. But sometimes something happens, like one of those little bugs doesn't get fed enough of whatever it is they need to stay alive. When that happens—they starve and die.

"So, when you're homesick, or when you get teased and you feel that ache inside—that's the insects dying— one by one. You've only got so many insects in you, boy. Can't let too many of 'em die. If that ever happens, you'll end up as numb as a bucket and as cold as a blizzard."

Jeffery figured Grandpa was about the smartest alive.

Outside, after breakfast, Jeffery resumed his station by the side of the road, his head turned to the crest of the hill. He knew his mother's car would come barreling over the ridge any minute now. She'd slam on her brakes and throw open the back door while the car was still moving. They wouldn't even bother to pack his clothes—they'd just buy him new ones later on.

The back door would close automatically by an unseen hand and Jeff would bounce up and down in the back seat, holding onto the safety-cord that ran along in front of him as the car picked up speed and fishtailed in the deep snow. Looking up he'd see soft auburn hair that flipped whenever his mother moved her head. Then he'd stop bouncing for a minute and turn in his seat, resting on his knees, and wipe the moisture from the rear window. The glass would squeak as his skin pushed the condensation aside and he'd watch his father's farmhouse shrink in the distance.

With his back still turned, his mother would watch him in the rear view mirror" and say "I love you, dumpling," and he'd reply, "I love you too, mom." Then he'd turn back around and sit quietly in his seat , his hands folded in his lap—his fingers intertwined to show his mother he was being a good boy.

"Where're we goin', mom?" he'd ask. "We're going to a special place, my baby," she'd answer. "Okay." Jeff's legs would be too short to touch the floor so he'd kick his legs and watch his shoelaces fly about like brown spaghetti. "What's the place called?" he'd ask. He'd look into the rear view mirror and his mother's eyes would leave the road to meet his. "It's a place called 'home'."

Jeffery stopped himself from pretending and looked back at the farmhouse. He could see Aunt Jackie's face in the window, but through the falling snow he couldn't tell if she was watching him or the frozen crystals falling from the pale gray sky.

Jackie slipped off her shoes and rubbed the balls of her feet on the furnace vent. The warm air blew up her legs, lifting the front of her dress in billowy ripples.

"What's he doing?" Jeffery's father was leaning against the door frame lighting his pipe and admiring Jackie's legs. Pete puffed on his pipe and the scent of burning cherries filled the room.

"He's just standing there."

"He's gonna catch pneumonia."

"He'll be all right. I made sure he bundled up good."

Pete puffed on his pipe. "Well, just the same, it'.s not normal."

"He misses his Momma."

Pete grunted and pulled on a plaid wool coat. "He's been standing out there all day, every day, now. Ever since he got here."

"I know."

Jackie heard Pete leave by the back door and climb into his truck. She found a cigarette and studied the drugstore calendar while she fumbled with the matches. January 1947.

When she returned to her chair by the window, she saw Pete's truck disappearing over the ridge. "Every day, now," she whispered. "Every day for twenty- seven days."

Jackie thanked God she wasn't a religious woman.

If she had been, the guilt would have consumed her and Jeffery might still be living with his mother. Kidnapping her younger sister's son wasn't her most noble deed, but like it or not, Joanie didn't deserve him.

Besides, she thought, Joanie didn't want Pete and gave him up. Now, Joanie had herself a new husband—a smart-ass Jew-boy from Back East. She could always have more children if she wanted.

Jeffery was all Pete had. Jeffery belonged to Pete. She tugged hard on her cigarette and pulled the smoke deep into her lungs.

That bitch always thought she was better than me, Jackie thought. But now, I've got Pete—and, I've got Jeffery, too. So, now who's laughing?

When they were kids, she and Joanie and Irma and their youngest sister, Doris, would sneak off during lunch, duck under the fence, then run into the fields of Garden Meadows. They hid amongst the trees and bushes, far from the eyes in the school yard.

Doris always brought along her rag doll, Missy, and kept her clutched close to her bosom. With her free hand, she'd ward off branches that snapped back in her face as she attempted to keep pace with her older sisters.

"Wait for me," she cried.

"Hurry it up slow poke," Jackie called over her shoulder as she side-stepped a gofer hole and leaped over another. Then, with a sudden burst of energy, Jackie hiked her dress up to her waist, dug her heels into the soft earth and sprinted across the field to the woods.

While Joanie stopped to wait for Doris, she watched Jackie high-tail it across the field, her tight rump straining against her worn yellow panties. "That girl has no shame," she remarked.

"Who doesn't?" asked Doris, finally catching up.

"Your not-so-bright sister," Joanie replied. She took Doris's hand and hurried her along. In their haste, Doris lost her grip on Missy. She quickly let go of Joanie and ran back to retrieve her companion. Joanie broke stride and waited for Doris to rejoin her, and soon they were running across the field together.

"I got an 'A' in Arithmetic," Doris boasted, out of breath. "That's because you're a clever little girl."

Jackie entered the woods and leaned against a tree while she caught her breath. Joanie and Doris had yet to catch up, which was fine with her—she couldn't tolerate either one of them.

She drank in the fresh air, her lungs thirsty for oxygen. The air was uncomfortably warm and overly fragrant from the thick blanket of decaying fir and spruce needles that covered the forest floor. The treetops were almost impossible to see, hidden by hundreds of branches and clusters of cones.

Jackie felt the need to empty her bladder. Without giving the matter a second thought, she pulled down her panties and squatted. Pee immediately drizzled onto the ground between her legs

"What'cha doin', Jackie?" Doris asked, entering the woods, still clutching her doll.

"What's it look like, turd-brain," Jackie answered, annoyed by her obnoxious little sister.

"That's a cruel thing to say," said Joanie, releasing Doris. "Tough-titty," said Jackie, looking off.

"I got an 'A' in Arithmetic," Doris bragged.

"Well, isn't that just grand," Jackie replied. "What'd you get?" asked Doris.

"My tests happen to be a little more difficult than a first-graders."

The urine continued to drain from Jackie's body, making a puddle where the earth refused to absorb any more.

"Joanie, what'd you get?" Doris asked.

"A 'B'," Joanie answered matter-of-factly.

"Well, bully for you, too,'" said Jackie, wrinkling her nose.

"Where's Irma?" Joanie asked, peering into the woods.

"I wouldn't know. Haven't seen hide nor hair of her."

"No, but we can see hide and hair of yours," said Doris, giggling and pointing between Jackie's legs.

"That's just what I expect a retarded little girl to say."

"I am not!"

"Yes, you are. Extremely retarded."

"I am not!" Doris shouted again, her face turning red, her eyes filling with tears.

Jackie nodded 'yes' and mouthed, "oh-yes-you-are."

Doris screamed—a high, shrill, ear-piercing scream that echoed across the wooded tract and bounced off the giant trunks of the towering pines.

"Shut up, you little brat, or we'll get caught." Jackie warned. She had nearly finished peeing, it was just dribbling out of her now, but the puddle had grown considerably, engulfing her shoes and turning the soil to mud.

"I hope we do get caught!" Doris hollered.

Realizing the only way to stop the confrontation was to separate the two, Joanie nudged Doris. "Come on," she said, leading her away. "Let's see if we can find Irma."

"Wait for me," Jackie whined.

"Forget you. We don't need your kind around us anyway," Doris sneered.

"What's that supposed to mean?" Jackie asked, flexing her sphincter and shooting one final stream. Her feet tingled and she was sure her legs were felling asleep.

"You figure it out smarty-pants. You think you know everything. Come on, Doris," Joanie insisted.

"Johnny Taylor says you're as loose as a goose and horny as a toad," Doris hollered at her, then turned to Joanie. "What's that mean—loose as a goose?"

"It doesn't mean anything, come on," said Joanie, taking Doris's hand and pulling her away.

"Johnny Taylor said I'm loose? How the hell would he know? He just wishes." Jackie put her weight on her heels and started to get up.

"Yes," said Doris, "he says you're loose and for a nickel you let the boys stick their fingers in you."

Jackie had a good mind to slap her little sister silly. She stood suddenly and slipped in the mud. It happened so fast she had no time to react. She splashed into the puddle, landing hard on her bottom.

"Shit!" she shouted, struggling to her feet. "Come back here!" But, Joanie and Doris ignored her, both covering their mouths to hide their laughter as they disappeared into the woods.

"Yeah, real funny," she shouted after them.

Jackie was careful to keep her dress pulled tight around her waist. If she got her dress dirty she knew she'd have hell to pay. Her panties were still stretched between her knees and splattered with mud. "Damn!"

She braced against a tree and slipped out of her soiled underwear. "Miserable little brat," she muttered. With her dress bunched in one hand and her underwear held aloft in the other, she pushed herself away from the tree and started for the watering hole.

Two classmates, Peter Nole and Johnny Taylor had also snuck off during lunch time. They took a different route, but were within shouting distance of Jackie's accident.

"Sounds like one of the Worthy girls," said Pete. "If you ask me."

They were sharing a smoke and happy to be away from the claustrophobic one-room schoolhouse when they heard Doris scream.

"No doubt," Johnny Taylor agreed and handed the the corncob pipe back to Pete.

They were following a trail partly hidden by dense undergrowth. It was a warm day and the woods were teeming with knats that hovered in swarms. Mosquitoes too, were present, but they were mostly loners, seeking out meals from warm bodies.

It was the season of young caterpillars. Green and yellow striped ones toppled to the ground after inching their way out of massive webs that draped the branches overhead. So thick were the nests, they formed their own silky clouds.

"I saw Merda looking at you in class today,"

"Yeah—right," Pete remarked, skeptically.

"She was."

"She's cross-eyed—how the hell could you tell?"

"Well," said Johnny, taking another puff and passing the pipe back. "One of her eyes was looking at you." He grinned. "On the other hand, she could have been looking at me—I'm not sure now."

"You know, Johnny," said Pete, chewing on the pipe's stem. "Sometimes you make me want to fart."

"Prove it."

Pete did and they both snickered. "I dare you to do that in class when Mrs. McCloud's back is turned."

"Yeah—right."

They stopped and sat on a large rock. The scent of old pine lay heavy. "You weren't in class yesterday. How come?" Pete asked.

"Chores. There was too much to do. One of our cows had a calf."

"Fun."

"You should have been there."

"No Thanks," said Pete.

"It was a real mess. She was breach and it took hours. Christ, my dad had his arms up that cow all the way to his shoulders trying to turn her."

"Did he succeed?"

"Eventually."

The sun was straining to break through the forest's canopy, when Johnny spied a movement amongst the trees. A figure was moving swiftly in the direction of the watering hole.

"You'll never guess what I just saw," said Johnny.

"You're right I probably never will."

"I just Jackie Worthy walking through the woods bare-assed naked."

"Yeah—right."

"I'm serious," Johnny insisted, lowering his voice to a whisper.

Pete rolled his eyes and began refilling the pipe. "Bare-assed naked, huh?"

Johnny stopped Pete by placing his hand over the bowl. "Pete, I'm telling you the honest to God's truth. I just saw Jackie's bare ass."

Johnny wasn't kidding. They grinned at one another and lit out in pursuit.

Stalking Jackie became a great adventure. While Johnny led the way, rushing ahead, crouching when he thought he had to, Pete held up the rear, staying well hidden and peeking around trees. If and when the coast was clear, Johnny motioned Pete forward. It wasn't long before they had her well within view.

"Holy shit," Pete whispered as he caught sight of her, "I don't believe it."

Jackie kept to the trail, moving quickly, swooshing away mosquitoes that attempted to hop a ride on her bare bottom.

Some mosquitoes, the ones that hummed an octave higher, the females, were driven by pregnant hunger and were careless to distraction. They met their death swiftly with a swat or were knocked senseless by Jackie's yellow panties.

One mosquito in particular stayed aloft, hovering nearby— observing. Perhaps it sensed danger, or perhaps the quest for survival requires no thought. Whatever small bit of brain matter it did possess, however primitive and ancient, told the insect to hold back.

It hovered around Jackie's legs, watching her warm body move. Through its own perception of time, Jackie moved slowly, passing through space with the grace of a cumulous cloud. Each stretch or contraction of thigh muscle was a creeping wave upon a sea of skin. Short dark fibers poked from between Jackie's thighs and a scent wafted into the warm spring air that was especially inviting.

The mosquito moved in closer, extending its long delicate legs, gingerly touching the hairs, testing for sensitivity. There was no response from Jackie; she continued to creep in slow-motion, a young, pungent mammal—fresh and ripe and redolent. The insect tensed its legs. When the moment was right, the mosquito lit gently, the tips of her appendages sticking to a shaft of hair.

Constantly on guard, prepared to flee if the swinging yellow object came too near, the mosquito waited. It watched her thighs pass to and fro, and was aware of the slow passage of time. Eventually, very gradually, the mosquito relaxed and waited, balancing without effort upon a single shaft of hair as Jackie journeyed on.

Pete caught up with Johnny and they hid behind the same tree.

"She's definitely on her way to the watering hole," Pete whispered as he watched Jackie's nakedness sway from side to side.

"Definitely," Johnny whispered in agreement. They followed at a distance, weaving between trees and crouching behind bushes for camouflage.

Jackie walked quickly down the beaten path, determined to reach the watering hole before she had to head back to the school yard. A twig snapped behind her and she stopped to look back, seeing only the alameda and the shadows of the trees falling over the path. Though she was sure she sensed a presence, Jackie shrugged and turned back to the trail and continued on.

Unlike her sisters, Jackie's sexual awareness began when she was very young. Most of her sisters would grow to consider the sexual act to be an unfortunate situation necessary to hold onto a marriage. Most would experience less than a dozen orgasms in their lifetime—only engaging in the act for procreation or the insistent demands of their husbands. They would consider their husband's sexual appetites to be perverse, while fearing the male penis and performing fellatio only in the dark.

When she arrived at the shore of the pond, she tossed her panties into the water and let them float. She squatted near the water's edge and slapped the water on her bottom, rubbing the mud away with her hands. She worked quickly and deliberately.

She pulled her panties from the water and used then to wipe down her legs. When she was done she rinsed the panties and laid then on a boulder by the waters's edge to dry.

Sensing the sojourn had ended, and satisfied that it was safe to proceed, the mosquito prepared to extract a meal.

Jackie became aware of a light tingling, then a sharp sting. She rubbed her pubis to quiet the interruption and the mosquito was smeared from this world into the next.

She had nothing to dry off with, so she stood in the sun. The heat felt good where it warmed her skin. She billowed her dress to make a light breeze and waited for her body to dry.

Johnny and Pete watched at a safe distance, out of sight, and marveled at the roundness of Jackie's rear.

"She's got a great ass," Pete whispered.

'You can say that again," Johnny agreed.

"She's got a great ass," Pete whispered again and they both laughed quietly.

"Maybe you should asked her if she way to play carnival," Johnny suggested unable to wipe the grin from his face.

"How do you play that?"

"Ask her to sit on your face, so you can guess her weight." They both chuckled a little louder than intended.

"You know, you boys can quit hiding behind that tree and come out anytime, now. I know you're there."

Johnny and Pete exchanged looks, then grinning, stepped away from the shadow of a large pine. Jackie kept her back turned, still billowing her dress in the spring air.

"It's getting late," said Pete. "Maybe we should all be getting back before we're marked tardy." He could feel his heart pounding,

"Yeah," Johnny agreed. "We probably should be heading back.

Jackie turned her head and for the first time looked at the two boys. "That's right Johnny—maybe you should go back."

Maybe it was her voice, or perhaps it was the intensity of her eyes that made Johnny's ears hot. Jackie possessed a presence. She was not easily intimidated. She had always known what she wanted and had the tenacity and patience to get it—regardless of the length of time or the pain necessary to acquire it.

Johnny took one step backwards and looked sheepishly at the ground. Jackie stopped billowing her dress and let it fall over her derriere.

"Well, see ya," said Johnny, turning to go.

"Yeah—right," said Peter, preparing to follow.

"Pete?" Jackie said, her voice stern, but lilting. She turned and faced him. "Nobody excused you."

"Catch ya later," Johnny said, making a hasty retreat. "Don't want to be late," he called over his shoulder.

Pete watched his friend hurry along the trail, then melt into the bosky green of the dense forest. He turned back to Jackie.

"Looks like it's you and me," said Jackie. She watched Pete shrug, "Aren"t you afraid of being marked tardy?"

"Nope," Pete answered.

"Why not?"

Pete shrugged. "My grades are so bad right now it don't make much difference anyhow.".

Jackie smiled and sat down on the rock by her panties. "They're still a little damp," she said, feeling them. "Would you mind helping me?" she asked.

Pete shrugged again. "What do you want me to do?"

"Come here," she said.

Pete hesitated. Jackie beckoned with her hand. The air that moved through the trees had a light chill to it, and though Jackie sat in the sun, the breeze made her nipples hard. "Yeah—right," said Pete and moved forward.

It was almost as if invisible strings were pulling him to the rock. Anticipation of the unknown made his head itch. He felt sweat dripping from his underarms, running down his sides and into his pants.

"Come on slow-poke. We haven't got all day," Jackie jeered.

Pete somehow managed to bridge the glade without stumbling, then stood before Jackie, still uncertain of what to do with his hands. He suspected he looked like a monkey when he let his arms dangle at his sides. He considered locking his fingers behind his head, but knew that would look ridiculous. So, he removed his hands from his back pockets and hooked his thumbs into his belt loops.

"You're not afraid of me, are you?" Jackie teased.

"Heck no," Pete insisted. "I ain't ascared."

"Good, " she said, handing her panties to him. "Then you can help me put these on."

The underwear that had been put into Pete's hands were warm from the sun, but still damp from the rinse. They were pale yellow cotton with off-white lace sewn into the elastic waist band. Pete instinctively examined the crotch, imagining what might fill them when they were worn.

Jackie watched. She'd had a crush on Pete since she was twelve. Pete was two grades ahead and never gave her a second glance. Since then, her body had developed. Now, at the age of sixteen, she knew she was appealing. While other girls her age might castigate or ignore a young man's advances, Jackie was not about to dismiss such a perfect opportunity.

There are moments when awareness becomes overwhelming— times when the present sharpens to an enlightened perception. These epochal memories become chemically carved, permanently etched into the folds of the mind and resurface unsummoned throughout the whole of our lives.

A decade later, sitting in a chair near the window, running her bare feet over the surface of the furnace vent, letting the hot air blow up the back of her legs, Jackie relived the moment. She gazed out the window. And, though her eyes watched Jeffery as he stood in the cold and the falling snow, Jackie was seeing that spring day ten years before.

The phone rang, and Jackie traveled back to the present, her face warm from the memory. The phone rang again and Jackie left her chair by the window and traveled to the kitchen to answer. It was Doris.

She only half listened as she touched herself through the fabric of her dress. "She's comin' back!" Doris was on the verge of panic.

"Who is?" Jackie asked, fumbling for a cigarette and striking a match on the window sill.

"Joanie!" Doris shrieked into the phone. "I just got a telegram. She wired me from Denver."

"What's she doin' in Denver?" Jackie asked, not really caring where the hell the telegram came from.

"She's taking a train; there was a stopover in Denver. Jackie, what in God's name are we going to do?"

"Well, I don't know about you, Doris," said Jackie flicking her ash on the floor. "But I was thinking of taking a nap this afternoon." She gently placed her big toe on the ash and felt it collapse into a fine powder.

"Jackie! You kidnapped her son!"

"You helped." Jackie couldn't help but smile as she listened to the panic-stricken voice of her little sister.

"Helped! Are you crazy? I really think you are out of your mind, Jackie. I always knew there was something terribly wrong with you. And now, we might all be going to jail because of your stupid hair-brained schemes!"

"Relax, Doris. You haven't got anything to worry about. You didn't do anything wrong."

"Didn't do anything wrong? I was an idiot to ever listen to you—for ever believing you. You lied to me!"

Jackie pulled on the cigarette. "Take a pill, Doris. You're having a conniption fit." She smiled when she heard Doris slam the receiver down, breaking the connection.

"Shit," said Doris looking at the phone. "Shit, shit, shit."

"Shit," came a small voice from behind her. Doris turned in her chair. Her two-year-old daughter, Nellie, sat on the floor in the kitchen, straddling an overturned sauce pan, a wooden spoon poised in one hand.

"Don't say 'shit'," Doris admonished.

"Shit," Nellie repeated and banged the bottom of the sauce pan.

"Nellie, if you say 'shit' one more time, I'm gonna paddle your butt. Do you understand me?" Doris warned, raising her voice.

"Butt," said Nellie and smiled. She banged the bottom of the sauce pan twice. Doris shook her head and suppressed a smile. She laid her head in her arms and closed her eyes.

How the hell she ever got mixed up in this mess she'd never know. It all began so innocently. She was just going to watch Jeffery for two weeks. Two weeks, while Joanie and her new husband traveled back East to find a place to live.

Why the hell Joanie ever thought she could make a new life in New York City was beyond Doris. She'd heard about those big cities—the streets choked with traffic, the millions of people milling about the thousands of garbage strewn sidewalks. Nobody in their right mind could live like that.

She didn't hold it against Joanie for marrying a Jew. Hell, you can't help who you fall in love with—it just happens. It's like being born or dying—you have no control over it. It's just a part of nature. But, what Joanie ever saw in that man was beyond her.

Sure he was handsome—tall, broad-shouldered, olive-skinned. His black wavy hair was brushed straight back off his forehead, and he sported a pencil-thin mustache. He walked with a cock-sure gait and when he smiled he kind of reminded her of Clark Gable. Only, it didn't help matters that his head was so jammed full of facts.

Hank Olstein was so smart he made everyone around him uncomfortable. He didn't know how to be sociable. He was incapable of making small talk. He knew nothing about farming, but insinuated it required little, if any, intelligence and anyone with half a brain could do it. He was more interested in politics and history and talking about books written by "fine" authors. Marrying a man with a wealth of knowledge was definitely a mistake. He intimidated everyone except Joanie.

It hadn't been all that long ago that Joanie and Hank had met—a little over two years now, Doris thought. It was a Saturday night . . .

 . . . The USO Club was packed to the gills. They couldn't have squeezed another person onto that dance floor if they'd used a crowbar. Sailors and soldiers from every branch of the service lined the walls talking, smoking, watching and flirting.

The cloud of tobacco smoke that hovered near the ceiling grew heavier and thicker through time. Gradually it fell until it's belly rested on the heads of the dancers. It moved—convulsing and swirling in time to the gyrations on the dance floor—the dancer's heads dived in and out of the descending cloud like they were bobbing for apples.

Joanie and her sisters, Iris and Jackie, arrived while the band was playing its rendition of Glenn Miller's "In The Mood." Jackie headed straight for the dance floor and began swaying to the music. It wasn't long before she was joined by a soldier.

Iris and Joanie headed for the ladies' room. While walking by some officers assigned to the Army Air Force, one soldier tried to catch Joanie's eye. But, she didn't notice—her attention was on the band and the dancers. "Hey!" he shouted over the din.

Joanie stopped and looked at him. "Hey, yourself," she said, smiling. She was wearing a calf-high, light blue dress under a tight-fitting, dark blue vest and a scarf, rolled to reveal her long neck.

"You doin' anything right now?" He asked, raising his thick eyebrows.

I'll be damned if he doesn't hold a striking resemblance to Clark Gable, thought Joanie. "No," she replied, intrigued.

"Good," he said, taking the cigarette from his lips and tossing it to the floor. "Step on that, then, will ya?"

Joanie wasn't sure if she had just been insulted, or if she was dealing with an unusual sense of humor. It was a strange way of flirting. But, she was still intrigued by his mien. She snuffed the butt with her toe.

"Will there by anything else, sir?" She asked, feigned a curtsey.

The soldier laughed. "I like you," he said.

"I suppose I should be thrilled," Joanie countered, sizing him up. He was leaning with his back against the wall with all the arrogance of a spoiled child. He was tall and good-looking. His demeanor suggested that he might even have a brain. "You sound like you're from Back East somewhere."

"New York," he said flatly.

"New York," she said pretending to be thinking. "Nope. Never heard of it. Is it big?"

"The biggest," he said, sliding a fresh cigarette between his lips. "What do you say we get out of here—go have a cup of coffee somewhere, or something?" he offered.

"Or something?" Joanie asked, suspiciously.

"Just coffee. Is there a place that serves coffee around here?"

"The Coach House—across the street."

Joanie fold her arms. "I thought you were going to ask me to dance."

"Dance?"

Joanie nodded and smiled coquetishly.

"Hell, kid—I don't know how to dance."

"Oh," said Joanie, a little disappointed. "Then, why did you come here?"

"Why, to meet you, of course," he said, pushing away from the wall and offering Joanie his arm. "Shall we?"

She hesitated. I don't even know this guy, she thought. I'm not going to go traipsing off with the first soldier that comes out with a good come-on line. Though, I have to admit, I am tempted.

Joanie turned and watched the dance floor. The trombones blared and the temperatures rose as the band hit the downbeat of a new song. Someone knew how to make a clarinet sing; its sweetness flowed effortlessly through the smoke-filled air, swimming into her ears until chills ran up her back and goose bumps popped out on her arms. Her toes were tapping and her body was moving without being conscious of the music's pull. On the outer fringes of the dance floor, Jackie now had two soldiers dancing with her. Her record was five. Joanie never could figure out why men were so attracted to her.

"Hello . . . excuse me," said the soldier, tapping Joanie on the shoulder. "Are we still occupying the same dimension?" Joanie turned back and faced the soldier. "Look, you don't have to be afraid of me, kid," he said drawing close so he didn't have to yell. "I'm not going to bite you." Joanie grinned. "Unless you want me to," he added. Joanie laughed.

"So, we're just going for a cup of coffee, right?"

"That's right," he said leading her out the door. "We're going to drink some coffee, smoke some cigarettes, and I'm going to tell you what a swell guy I am and you're going to fall head over heals, madly in love with me."

"Oh, is that so," Joanie countered. Boy can this guy gab, she thought.

"A-huh. And then, at the end of the evening, I'm going to ask you to wait for me while I go overseas and bomb the hell out of the Germans." He secured his grip on her arm for protection as they descended the stairs that led to the outdoors.

"And, how will I respond?"

"Oh, naturally you'll agree."

"I will?"

"Then, of course, later on, I'll ask you to marry me."

He stopped to light a cigarette and offered her one. She accepted. "And, naturally, you'll agree."

"Naturally," she said, playing along.

They sat in a booth by the window in the the Coach House Diner and drank coffee from oversized cups. He said his name was Henry Olstein but his friends called him "Hank." He graduated from high school when he was sixteen and already had three years of college under his belt by the time the U.S. entered the war. He was nineteen when he joined the Army, They made him on officer and taught him to fly.

He was raised in a family of strict Jewish Orthodoxy on the second floor of a walk-up in New Rochelle, New York during the years of the depression. His father was an upholsterer who learned his craft in Odessa, Russia. In 1917, in the midst of the Communist revolution, when crowds rioted in the streets and the snow was spattered with blood, his father, Leo Olstein, decided Communism was a social experiment he wanted no part of. He left his beloved country and journeyed west.

It was raining the day his ship, an Italian freighter, pulled into the harbor of New York City. Along with a thousand other immigrants, Leo stood in line on Ellis Island and waited for permission to begin his new life. His father before him had immigrated to Russia from Holland—converting his Dutch name into Russian. Now, Leo was attempting to translate his Russian name into something pronounceable in English. A clerk performed the task for him with the stroke of a pen. "Your new name is 'Olstein'," he was told.

"Olstein," Leo thought to himself. He rolled the word around in his head. Olstein—what the hell does that mean? Where he came from, surnames had meaning—they stood for something. "Ach," he decided aloud, "what do I care? A name's a name. Nobody cares what your name is—this is America."

He married Sara, a woman twenty years his junior and fathered two children—a girl, Franny and some years later, in 1924, a son, Henry. As Henry grew, Leo attempted to enlighten his son with the fruits of his wisdom. He compared anything and everything to Russia before the revolution.

"Leo!" Henry called to his father one day. Henry ran across the street and stopped at his father's side—a little out of breath.

"Home from shul so soon?" Leo asked in Yiddish.

"Sure," Henry said. "I got home early." He pointed across the street to a vacant lot, where a building used to stand. "I took a shortcut," Henry said proudly.

"Henry, let me tell you something," Leo said, shaking his finger disapprovingly. "Never take shortcuts."

"But, I got home sooner."

"I don't care" said Leo raising his voice. "Never take shortcuts for anything—for any reason.

His friends thought it strange that Henry called his father "Leo," while they referred to their fathers as Daddy or Papa.

One evening while eating dinner when Henry was eight, he decided to call his father "Dad", thinking the new endearment would be something his father might appreciate. He was promptly backhanded and knocked off his chair onto the floor.

"My name is Leo," he shouted. "Don't ever forget that!"

Henry never did. It wasn't until Leo was on his deathbed, some sixty years later, that he would suddenly change his mind. "Henry," he would say—"I'm your father, can't you call me Dad?"

Leo was someone Hank would never quite figure out, though he never stopped trying. When it came to gift-giving during the Holidays,

Leo was stingy to the point of embarrassment—at least when it came to his own son. On the other hand, he never thought twice about paying for the passage of his three sisters and their husbands to come to America. He even helped them get set up in businesses.

On Fridays, Leo insisted that everybody should come to his home for the Sabbath. Of course, this made more work for his wife, Sara, but she didn't mind. After all, it was the Sabbath—a chance to dignify her home, a chance for the woman to lead the prayers, a time to display the short silver candleholders and prepare something very special for dinner. For hours they would sit around the table to eat and laugh—Yiddish, the only spoken language.

"Leo," said one of his sisters after dinner one evening, "I simply love this table."

Leo wrinkled his forehead and looked at his sister like she was crazy. "Table? What table?"

"This table" said his sister., tapping the table top.

Leo looked at the table. He turned in his chair and looked under the table then sat upright. "You love this table?"

"Yes," she declared. "it's absolutely beautiful."

Leo shrugged. "It's just a table."

"It's eloquent," she said, her eyes alive.

"Eloquent? It's a table."

"Where did you get it? I want Harry to buy one of me, just like it." Harry nodded.

"Just s soon as we can afford it."

"Where did I get it?" said Leo rubbing his chin. He always rubbed his chin when he was thinking. "It's just a table, Who remembers such things?"

"Think Leo. Try to remember. I want one just like it."

"You want a table just like this table?" His sister nodded in agreement.

"Here," said Leo, getting out of his seat. "Take the table. Henry, help your mother clear these dishes off."

"But, Leo . . ." his sister protested.

"But, nothing," Leo said. "Take the table. Take it home with you."

"But, Leo," Sara objected. "That's out table. I like our table. What will we eat off of?"

"I'll buy you another tablet tomorrow. Tomorrow I'll buy you another table just like this one. All right?"

Eventually, the table was replaced, but it was nothing like the original and it wasn't for a couple of weeks. In the meantime, they balanced trays in their laps and Sara fumed.

During the two and half hours they sat and talked, Henry and Joanie consumed ten cups of coffee and a pack of Lucky Strikes. They shared a slice of tepid apple pie smothered in hot cinnamon sauce and a scoop of vanilla ice cream on the side. Finally, Hank walked Joanie back to the USO Club where he demonstrated convincingly why he was considered to be such a poor dancer.

After the band played their last song and was busy packing up their instruments, Hank slipped a note with his APO number into Joanie's hand and kissed her on the lips. A blanket of warmth wrapped around her and Joanie fought to keep her balance. The energy that flowed from him was electric, and though she didn't think it possible, she was aroused by that single kiss. He asked her to write if she got the notion. She smiled and promised she would.

As the bus was pulling out, Hank leaned out the window and flicked his cigarette into the air. It sailed several yards in a long arc before plummeting to earth.

"Save that for me, will ya?" he shouted. "I'll be back!"

Joanie laughed aloud at the memory. Her sisters were sitting near her on the front porch, each, for the moment, lost in her own thoughts. There was a moment of quiet while the sisters refrained from visiting and listened to the night. The waxing moon was a sliver crescent, the stars exceptionally bright.

"Pardon me, ladies. Might the be the Worthy house?"

Joanie nearly jumped out of her seat. "Hank!"

Henry Olstein was leaning on the picket fence. He smiled and removed his service cap. "Evening ladies " he said nodding to Joanie and her sisters.

"What are you doing here?" Joanie asked, shaken.

Hank opened the gate and sauntered up the walkway, donning his service cap again. "I think a better choice might have been . . . Hank, how good to see you, won't you please join us?"

"Hank," Jackie said, sliding off of the porch railing and rising to her feet. "How good it is to see you," She smiled broadly. "Won't you please join us?" She gestured to the poor swing, Joanie was sure she could hear Jackie purring.

"Thank you. So kind of you to offer. And, you are?" he asked settling onto the swing next to Doris and crossing his legs. "Doris," said Doris, flushing.

"Doris," Hank repeated, shaking her hand. "I'm pleased to meet you. You're much prettier than Joanie tells. She said you were a skinny little runt."

"I did not! He's just pulling your leg, Doris. I said nothing of the sort ."

"She's right, I'm only kidding. She said you were the prettiest of the lot, and I believe her." Doris flushed deeper, but was flattered.

"I'm Jackie." said Jackie, joining them on the swing, leaving Joanie and Iris standing.

"Well," said Iris, putting out her cigarette, "I think it's time I turned in. Nice to have made your acquaintance Hank. Goodnight all."

"Goodnight," said Jackie, pretentiously pleasant. "Sweet dreams," she said, wiggling her fingers adieu. Iris rolled her eyes at Joanie as she brushed past and entered the house.

Joanie watched Jackie fall all over herself helping Hank light his cigarette. She laughed and giggled at every nuance or comment he made. Hank was evidently enjoying all of the attention.

"I hate to interrupt this fascinating conversation," said Joanie finally, "but, what are you doing here, anyway?"

"I live here," said Jackie. "She can be so forgetful sometimes," she confided to Henry and laughed. Joanie didn't find Jackie's humor the least bit amusing.

Henry sensed her awkwardness and rose. "I'm sorry," he said, facing her. "I came back because I . . . say," he said as if a thought had just occurred to him. "How do you get to Coeur d'Alene, Idaho from here?"

"Just stand in front of Warner's Drug Store. A bus will happen by sooner or later," Jackie responded from the swing.

"That sounds easy—come on," he said, taking Joanie's arm.

Joanie pulled her arm away. "Where are we going?"

"To Coeur d'Alene."

"Why would I want to go to Coeur d'Alene?"

"Because there's no waiting period in Idaho. We can get a marriage license and be married all in the same day."

Joanie felt like she'd been slapped. At first she wasn't sure she'd heard him right. But, she had. She looked back at Jackie and Doris. Both were shocked. Jackie's mouth had fallen open and remained that way. Joanie liked that. Henry was gazing into her eyes and wearing a smile.

"Are you asking me to marry you?" She asked, barely able to get the words out.

"Yes," he said, removing he service cap. " I am."

"But, we hardly even know each other."

"We've got the rest of our lives for that."

Joanie looked back at her sisters again, hesitating. Then, shrugging her shoulders while suppressing a grin, she took Hank's arm and started down the stairs. "Okay," she said with finality.

Jackie came to her feet. "Okay!? You mean you're just going to go off and marry him?"

Looking at Henry she replied, "I'd be a fool if I didn't. Watch Jeffery for me?" Doris nodded. "I'll be back by tomorrow—sometime—I think."

Doris shook her head and suppressed a smile. Nope, it hadn't been all that long ago that Joanie and Hank had met. About two years ago, she guessed. She laid her head in her arms and closed her eyes. How the hell she ever got mixed up in this mess she'd never know. It all began so innocently.

At the end of the war she agreed to watch Jeffery for a couple of weeks while Joanie and Hank went back East to find a place to live and Hank finished college. She looked at the phone; the receiver was still resting in its cradle. Just moments before, in a fit of frustration, she'd hung up on Jackie.

"She's coming, Jackie," Doris said to the phone. She looked out the window at the falling snow. "Joanie's coming and she's pissed."

Not far away, Jackie stood at the window. Through the falling snow she saw Jeffery's eyes lock on her's. Kids are resilient, she thought, they come back. Before long Jeffery would forget what his mother looked like; in a couple of years he won't even care.

Jackie snuffed her cigarette in a cracked ashtray and sat back down to watch the snow fall. She liked the way the warmth of the furnace air rose out of the floor vent and pushed past the back of her knees as it slid up her legs to her waist; it felt good.

I'll say one thing for him, Jackie thought, he's persistent. That kid's got a one track mind—standing out there everyday now for twenty-seven days. He hasn't done anything else. Doesn't play with his toys, doesn't talk about anything—just waits out there in the snow expecting at any moment to see his mother's car come barreling over that hill. He's got a one track mind alright, kinda like his father.

Pete came to Jackie every night and took her. She knew sex was all Pete wanted from her, but that was fine with Jackie—that's pretty much all she wanted from him. It wasn't like they shared any thoughts, or had any deep conversations. What the hell was there to talk about, anyway?

Jackie picked up a book and read for a moment, then took a pencil and crossed out a few words, replacing them above or in the margin with words of her own. She read a little further, then drew an "X" through an entire paragraph, deleting it. That's better, she thought. This Steinbeck fellow sure as hell don't know how to write.

Outside, Jeffery had turned away from the window and watched his father's truck pull way, disappearing over the ridge—the snow not yet settled where the tire tracks ran along the road. Then there was nothing. Nothing, but the knob of the hill where the fence disappeared over the ridge.

Instead of wondering why his father hadn't bothered to stop to say anything, Jeffery thought about grasshoppers and wondered how they spent the winter and what kind of homes they made in the snow. He knew what they did in the summer.

Last summer, when the war was over, he and his mother moved to Savannah, Georgia for a spell to wait for her new husband to return from Europe.

The summer was slow to warm up, but when it did, it turned into the hottest summer Savannah had encountered in nearly a century. The trees suffered terribly, most of their leaves withering and turning brown before dropping from the branches. The ground turned so hard and dry, that when it finally did rain, the water just sat in puddles as if the ground had forgotten how to swallow. Tiny lakes formed, which pleased the birds just fine—flocks of robins, sparrows and mockingbirds converged on the pools for joyous community baths. Their chirps and songs could be heard for miles.

The skies stayed clear and blue, but as the summer wore on, the heat climbed and the sky's color washed away, bleaching ashen. The hotter it got, the higher the grasshoppers jumped, till there were times when you couldn't look across the road without seeing their green wings clapping through the air.

When Jeffery wondered aloud why the grasshoppers jumped so much during the summer's heat, his mother's answer stuck to him like honey. "To keep from burning their little feet, hon'. Grasshoppers have feelings too."

But, that was then—five long years ago. Now, lying on his back with his fingers locked and supporting his head, Jeffery sighed and opened his eyes on the banks of Hangman's Creek.

Nellie was straddling the rope and gliding gracefully over the stream. She was talking, her voice no different than the babble of the water that painted the rocks with cold and wet and sparkling reflections of sun dots. Her words carried no meaning, but were bent with the inflection of candor.

She was suspended above and looking down as she spoke, her words making little sense—only now becoming audible, her knuckles white from squeezing the rope as she strained to make herself heard. Jeffery wasn't sure what she was blabbing on about, but, he knew one thing for certain, Nellie wasn't wearing any underwear. Beneath that pretty polka dot dress she was still as naked as a Jay bird.

Jeffery caught a word here, a sound there. Sometimes her mouth moved but the wind carried her words way, or the birds drowned out her voice. "Only I wasn't supposed to tell you"—Jeffery watched her polka dot dress billow with the movement of the air—"a secret." He studied the leaves above her head that sometimes fell into the creek and became little boats—"my mom is your mom's sister"—little boats that sailed away, becoming smaller, rounding the bend with the moving water and disappearing, never to return—"So, I know it's the truth."

Jeffery's eyes wandered back to Nellie who seemed to be weeping and mumbling words that became clearer with every gust of wind. "I'm sorry I said anything, Jeffery. But, if I had a brother I'd want to know."

Nellie stopped speaking, and Jeffery watched her as she drifted quietly over the creek. The rope groaned and creaked. She avoided his eyes and looked off into the woods. "Do you want to know his name?"

"Who's name?"

"Have you been listening to a single word I've said?"

Jeffery saw the streaks of dirt on her cheeks. "What are you bawling about?" he asked.

Nellie slapped the smudge of tears away. "I'm not bawling—screw you, you dumb bastard!"

Jeffery sighed. He'd be damned if he'd ever figure out what the hell went on inside a girl's head. Slowly the jumble of Nellie's words collated into a semblance of meaning. The shock of their implication was unnerving and Jeffery figured he probably misinterpreted something. But, just in case, he asked . . . "Did you say I had a brother?"

"Yeah. You were listening! Actually, he's your half-brother. His name's Benny. Benjamin, I guess. But, I've always called him Benny."

"You've seen him?"

"Sure, lots of times. First time we met, he did't even know we were related."

Jeffery was quiet. Considering. "Where was I?"

Nellie shrugged. "I don't know. Someplace else, I guess. I didn't think you were supposed to know."

It was too much to think about all at once. Too much to consider. Jeffery shifted his weight uneasily, pressing his legs together. Suddenly, he had to go. Bad.

"What's wrong with you? You got a bug up your butt or something?"

He tried ignoring her.

"You gotta go to the bathroom?" Nellie was still straddling the rope, her butt resting on the giant knot.

Jeffery nodded.

"So, go—why don't ya?"

Jeffery looked about sheepishly then wandered off to a spot out of sight and relieved himself . . .

. . . Jeffery shook his head and opened his eyes. He'd been remembering for a long time—thinking about the day before. He was sitting in a spot in the hay where the rising sun poured through the missing slats in the walls of the barn. Early in the morning he'd climbed to the highest loft where the ceiling beams supported a roof that hugged the sky.

It was time for breakfast and Irma had just rung the bell that hung on the back porch of their farmhouse. He could hear Duke barking and jumping up and down while running around Harold's legs as Harold made his way down the dusty drive for the kitchen.

I have a brother, Jeffery thought as he descended the ladder and left the loft. Benny. I've had a brother for five years and everybody's known, but nobody had enough guts to tell me until yesterday. Nobody, but Nellie. You gotta love that kid.

The heat of the sun hit Jeffery in the face as he left the coolness of the barn and stepped into the open air. Barney and Peabrain, the two roosters, were exhausted from attempting to kill one another and were now strutting back and forth in front of the chicken coop, ruffling their tail feathers and clucking.

Jeffery stuffed his hands into his pockets and kicked the cow turds that lay scattered in the dusty drive. He heard the screen door complain then slam as Harold disappeared into the house. The words of a lullaby whispered to his memory and the voice of his mother sang in his mind—

"You're a very special boy, Mister Man,

You're the only one for me in all the land.

If you ever need a hug,

Or you find your all alone,

Turn the toes of your boots

To a place called home

I grew up in Spokane, Washington, a large farming town poised on the banks of a fast moving river. Like most small boys, I was plagued by hobgoblins of my own invention and terrified by an overactive imagination that left me tear-streaked and exhausted.

My house rested on the crest of a small hill that belonged to a chain of hills—some rising above, others rolling below. Those above were wooded and green and dotted with houses that wound their way to the summit. Below, the hills ended abruptly at a gully rich in flora and small animals. Beyond that, the land leveled off and became the Spokane Valley.

In the spring, the rains came and washed the streets clean. I spent wet days next to the window, seated at a table, daydreaming. Rain ran down the pane leaving trails of micro-rivers that blurred the outside world, and I wondered why only cars had wipers.

On the day of the family picnic thunderheads rolled down out of Canada and hung ominously on the horizon.

The oversized tires of the rumbling Buick moved swiftly out of Spokane and headed South. I sat in the back seat behind my mother and watched my father drive. The upholstered seats were generously cushioned and the car's suspension was loose, so whenever the Buick hit a dip or a rise, the vehicle bounced and rocked gently for miles afterwards.

"I'll bet you a dog-turd against a donut it'll be raining before we get to Irma and Harold's," said Hank nodding to the threatening clouds. He downshifted at a corner and turned his attention back to the road.

"That would be the shits," Joanie responded. She was in the passenger seat holding her newborn daughter, Holly.

Between them on the seat, were two freshly baked pies. The scents of lemon meringue and caramel sugar bonded with the leather upholstery and cigarette smoke reminding me of the circus.

The circus—the excitement, the noise, the animals, the clowns and Grandpa. Grandpa Leo stood in the elephant tent holding my hand while I watched the animals lift hay to their mouths with their long gray trunks.

Last summer Grandpa Leo accompanied by Grandma Sara and my dad's sister, Franny, journeyed by train from New Rochelle, New York. For two weeks they complained about the drawbacks of living in a backward farming town in the Pacific Northwest and bragged at every opportunity about the marvels of New York.

On the morning of their first Saturday in Spokane, breakfast was served promptly at eight.

"Come on, Benny—sit down," my mom said, placing a platter of fried potatoes smothered in onions on the table.

Grandpa Leo sighed heavily as he scooted up to the table and surveyed the assortment of food. The spuds sat in a shallow pool of melted butter, steam scurried away then dissipated as it met the cooler air of the room. Definitely too hot to eat just now. A percolator piped coffee into its glass top, hiccuping in staccato burps. English muffins and dry toast rested in a warmer, covered by a towel. "Where's your yarmulke," asked Leo, eyeing me.

"My what?" I asked, settling into my seat.

"Your yarmulke," Leo insisted.

"What's that?" I asked, playing with my fork. Silence.

"He doesn't own a yarmulke," my dad said, avoiding his father's eyes while pulling his chair out from the table.

"Why not?" Grandma Sara asked.

"Because he's not a Jew," my dad said flatly.

"He's half a Jew—that's good enough for me," Leo announced, patting me on the back.

"What a yarmulke?" I asked.

Grandpa Leo pointed to the yarmulke resting on his pate. "What do you call this?"

"A beanie," I said, matter-of-factly.

"A beanie!?"

"Sure, But, it's missing the propeller."

My father laughed.

In defiance, Leo removed his yarmulke and planted it on my head. "There. Now you're a real Jew."

I thought about that. "Am I a whole Jew now?"

"No you're still half a Jew, but you're a good half a Jew," Leo responded with conviction.

I thought about this as I piled food onto my plate. "Which half of me is a Jew, Grandpa?"

Francis snickered while attempting to look serious.

"The best half," Leo said, then realized he may have insulted my mother and quickly regretted saying anything.

"So, what's the best half, the top half or the bottom half?"

"There is no 'best half'," said my dad, passing a platter. "Both have their good points."

"Do both halves have their bad points?"

"The Jew half doesn't have any bad points," said Leo then kicked himself for opening his mouth again.

I finished chewing, swallowed hard then washed the food down with a gulp of milk. "Hitler was half a Jew," I announced suddenly.

Leo dropped his fork onto his plate. "Where does this kid come up with these things?" he asked, overcome with astonishment. "Let me tell you something. Hitler was a son of a bitch."

"Leo, don't swear in front of the child," Grandma Sara admonished.

"So, who's swearing?" He looked around the table innocently. "He was a son of a bitch. That's a fact. That's not swearing. You want swearing I'll give you swearing. I've got plenty more words about Mr. Hitler."

"That won't be necessary." It was the first time my mother had spoken since joining the breakfast table. "Benny's well aware of how much Hitler is hated."

"Good!" said Leo patting Benny on the back again. "Sometimes hate can be good. As long as you hate the right thing." He watched his grandson eat for a moment. "You look good in my yarmulke. But, you need one that fits you. You want I should buy you one?" I nodded, my mouth full. "Okay, this I buy for you. A new one. One made in Jerusalem." He beamed then returned to his meal, digging into his plate with renewed vigor. He scooped up a forkful of fried potatoes and scrambled eggs and shoveled them into his mouth. "Do you know where Jerusalem is?"

I took a wild guess. "Idaho?"

But, Leo didn't hear me or chose to ignore my response. His mind was thinking way too far ahead. "Henry!" Leo said, almost shouting.

My dad looked up from his plate.

"Where's a Judaica Shop around here?"

My father shrugged. "I don't know, Leo. I"m not even sure Spokane has one."

Leo stopped eating and stared at his son. "You don't know?" My dad shook his head then resumed eating. "How can you call yourself a Jew and not know where a Judaica shop is? What kind of a Jew is that?" My dad didn't bother to reply; he reached for a slice of toast and spread jam on it. "You forget your heritage. You forget what it means to be a Jew."

"I haven't forgot. I just never think about it," my father lied. The truth was he never stopped thinking about it. But, he didn't feel he had to explain himself to his father anymore. He was all grown up now, he had a family of his own, he could conduct his household as he saw fit.

"How could you forget?" Leo asked, almost whispering. "You should never forget."

"An elephant never forgets," I piped in.

Leo was shaking with rage when he turned away from his son to look at me. "An elephant?"

"Sure. An elephant never forgets."

Leo resumed eating, but with less enthusiasm than before. "How do you know that?"

"I don't know," I shrugged. "It's just a common truth."

"A common truth?" Leo said, turning to his wife. "Sara, where does this boy come up with these things? Benny, you want a common truth?" I nodded. "An elephant has a brain the size of an acorn. But, it's not the size of your brain that counts, it's what's inside of it."

"Okay," I said, not really sure what Grandpa Leo was implying.

"You ever seen an elephant?"

"Sure," I said. "Plenty of 'em."

"Where?"

"In picture books."

"Hank, why don't you take your son to see the elephants?"

My father shrugged and looked out the window. How can you hate a man and love him at the same time? Such a paradox, he thought.

"What elephants?" I asked.

"You're not going to take your son to see the elephants?" Leo asked. "I have an arraignment on Monday. I have to get ready for it."

" 'Rainment—Shmainment, the boys never laid eyes on an elephant."

"What elephant?" I asked again.

"Maybe I should take you to see the elephants," Leo suggested, thinking out loud, "Would you like that?"

"Sure," I agreed. "When?"

"After breakfast. After you take a bath and comb your hair."

"Where's the elephants?"

"You'll see. Is that all right with you, Henry?"

My dad shrugged. "Fine," he mumbled.

"Joanie?" Leo asked, turning to my mother. My mom was unsure. She was afraid to let me out of her sight. "I won't let him out of my sight," said Leo, sensing her worry. "I'll be next to him all the time. I'll feed him lunch at the circus, too."

"The circus!" I shouted, dropping my fork and slipping out of my chair. "Holy cow," I said, rubbing my hands together with excitement. "We're going to the circus?"

My mother nodded and smiled. "I didn't know we had circus!"

"It passes through once a year, stays for about a week, then moves on to the next town," she said.

"Holy cow," I exclaimed again. "Can I wear your yarmulke to the circus, Grandpa?"

After breakfast and two hours before noon, our cab pulled up next to three giant tents erected on the outskirts of town. I held my grandfather's hand and kept one eye on the ground, sidestepping the piles of camel, horse and elephant dung. My shoes were new and a little too large. One shoe kept slipping off, and more than once I nearly stepped out of it. I watched my shoes grow dusty while I attempted to keep pace with Leo's long stride.

Leo's hands were thin and bony. His white hair was thinning, but his posture was that of a soldier's—his gait was quick and sure. Leo held his head high, rarely looking down, except when I stumbled or dawdled.

On our way to the big tent we passed through a corridor of sideshows. Leo slowed his pace when my attention was caught by something shocking or unusual. The barkers sang their chants and sold their tickets.

"You there, sonny!" I turned and was surprised to see a man in a black top hat and tuxedo pointing his finger at me. I stopped and squeezed my grandfather's hand for reassurance. "What did you eat for breakfast this morning, sonny?"

I looked around helplessly. Maybe he was talking to somebody else. Why does he want to know what I had for breakfast? The barker was looking directly at me; when he smiled his waxed mustache rose awkwardly at an angle. I looked up at my grandfather.

"Go ahead," Leo encouraged me. "Tell the man what you had for breakfast."

"Well," I began, "scrambled eggs, potatoes with onions, kosher salami sliced into squares . . . "

"Bruce, here, hasn't had breakfast yet," the barker proclaimed, gesturing to a stocky fellow standing next to him. Bruce wasn't wearing a shirt, and his pants were more like pajama bottoms, black, loose, and tied with a string. His arms were folded over his chest and his bald head had small dents in it. "Bruce, here, is really hungry. Aren't you Bruce?" Bruce nodded without smiling. "Can you guess what Bruce eats for breakfast?" I shook my head slowly. "Show him. Bruce!"

Bruce stepped forward, his bare feet stomping and shaking the makeshift wooden stage. He reached into a barrel and pulled out a light bulb. Looking directly at me, he opened his jaw wide and slid the light bulb into his mouth. He smiled with his eyes then slammed his mouth shut, crushing the light bulb with an audible explosion.

I felt my grandfather start, then hiss, "Jesus Christ."

"That's right folks, Bruce here eats glass for breakfast. He eats glass for lunch and dinner, too! Don't you, Bruce?" Bruce nodded as he chomped on the splinters, parting his lips to show the gathering crowd his teeth biting down upon the shards of shattered glass. "And, when he's not dining on glass, Bruce here likes to pound ten-penny nails into his head. Don't you, Bruce?" Bruce nodded and produced a large mallet.

I felt a tug as my grandfather pulled me quickly away.

"That man is eating glass, Grandpa," I said as Leo headed for the ticket booths.

"The man is an idiot," said Leo, reaching for is wallet.

"You wanna go back and watch him pound nails in his head?"

"No, thank you," Leo replied politely while pursing his lips in distaste.

I watched Leo pay for the tickets. "How come you said, 'Jesus Christ,' Grandpa?"

"I never said that," Leo denied, accepting his change and slipping the tickets into his jacket.

"Sure ya did. I heard ya. You said it when that guy was eating a light bulb."

"Come on," said Leon taking my hand again. "The show will be starting any minute now."

"If you're a Jew, how come you said 'Jesus Christ'?"

"I didn't mean to say it—it just came out naturally."

"Is it okay for a Jew to say 'Jesus Christ'?"

"Of course it is. Anybody can say it. It's okay," Leo assured me as we walked under an archway leading to the big tent.

"Do you even know who Jesus is?"

Leo stopped and knelt until he was level with me. "Of course I know who Jesus is. You think because I'm a Jew I don't know? Everybody knows."

"Do you hate Jesus?"

"Why would you ask such a thing?"

"Billy Lupo says that Jews killed Jesus, so I just wondered . . . "

"You tell Billy Lupo you can't be held responsible for what your ancestors did. Listen, Benny, let me tell you something. Even a Jew can admire a great man. We would be fools if we didn't. Did you know Jesus was a Jew?"

"I thought Jesus was a Christian."

"Jesus was a Jew—that's a fact."

"Did he wear a yarmulke?"

"Probably—or, at least a prayer shawl and phylacteries. Jesus was a good man. A smart man. Maybe the smartest man who ever lived. He's probably the most influential man since Moses. Just so you'll know," he said, dropping his voice, "I admire Jesus. He was the most important Jew ever born." Leo stood and took my hand.

"Let me tell you something, Benny," said Leo as we entered the tent and the scent of wild animals rushed to meet us. "They just don't make Jews like Jesus anymore."

Leo bought hot dogs, popcorn, cotton candy, and ice cream bars. We shared a warm bag of fresh roasted peanuts that was placed on the bench between us. The benches were hard; the air was stuffy, but neither of us minded because now the clowns were parading past waving and smiling.

Clowns rode on the backs of Shetland ponies and pedaled trikes; women in black mesh tights and tiny dresses rode the backs of elephants, camels and stallions—every one of them sporting feathers atop their heads and prancing in time to the music.

Bengal tigers growled and balanced on barrels to the crack of a whip, trapeze artists swung back and forth overhead performing gravity defying acts while a man carrying a long pole balanced upon a tight rope high above. Clowns shot themselves out of cannons and stuck firecrackers down one another's pants. Finally, two hours later, the last act was announced.

"Ladies and gentlemen, boys and girls, I direct your attention to the center ring," came the amplified voice. The lights dimmed and the crowd quieted. "The most death defying feat ever attempted."

The deep growl of a lion broke the stillness and the spotlight caught the cat's unexpected entrance into the center ring. The lion growled again then snapped his tail angrily as he stopped and surveyed the crowd.

"That lion's mad," I said.

"It's all an act," replied Leo. "He's supposed to act that way. What fun would it be if he acted tame?" Leo handed over a bag. "Here, you want some more popcorn?"

With the hand that held the hot dog, I took the bag and tried to balance it on my leg while the chocolate coating on the ice cream bar I held in my other hand turned soft and shiny in the warm tent.

The lion suddenly bounded to a raised platform and promptly sat. He seemed to be thinking for a moment, then chomped at the air with his teeth and shook his mane. He growled again.

"Ladies and gentlemen," the amplified voice announced happily, "Barnum and Bailey takes great pleasure in presenting—after years of safari in deepest, darkest Africa and fresh from his tour of the European continent . . . Rolando the Magnificent!"

The band struck up a glorious fanfare and I followed the spotlight as Rolando entered the tent. He bowed and tipped his safari hat while more lions raced past him to join the lone cat in the center ring. Rolando turned in a circle, greeting the crowd with his hat extended, his sequined cape billowing. A breeze blew through the tent and ruffled the canvass, temporarily relieving the crowd from the heat.

I discovered that balancing a bag of popcorn on your leg while watching a death defying act was no mean feat. Ice cream dribbled down the stick and onto my hand as the chocolate began to breakaway.

Rolando cracked his whip and all the lions except one hopped down from their perches and formed a circle around him, running follow-the-leader. He cracked his whip again and the lions turned and trotted in the other direction.

"Look how he controls those lions, Benny," Leo said with admiration.

I looked up and nodded while catching a sliver of chocolate with my teeth. I tried to catch a glimpse of the show, craning my neck as the remaining chocolate slid off the melting ice cream and fell into my lap.

Rolando snapped his whip again and the lions stopped in mid-stride. "Up—up!" he ordered and the lions sat up, balancing on their hind legs and waved their paws submissively. The crowd rewarded each feat with generous applause, whistles and cheers.

I tried to recover the chocolate from my pants before it melted, but I only managed to spread what was already there over a wider area. In the meantime, my shoe had slipped off my heel, but I was managing to keep it from falling onto the floor by balancing it on the end of my foot with my toes.

Rolando dropped his whip and removed a pistol from a side holster. He pretended to shoot the lions, firing his cap gun with great theatrical embellishments. The lions rolled over on their sides and played dead. Leo joined the crowd with hearty applause and a wide smile.

"Look at that, Benny!"

I attempted to see what was going on, but the crowd in front was constantly on their feet blocking my view. And besides, right now the important thing was to keep the bag of popcorn from falling to the floor. The ice cream, no longer protected from the elements by the chocolate coating, was now vulnerable to the warm air. It ran down the stick onto my hand, then followed my arm to my elbow where it dripped into the bag of warm peanuts. I tried to consume as much as possible, getting as much on my chin as I did my mouth.

"Up—up!" Rolando shouted. The lions scrambled to their feet then ran to their perches.

Then there was a hush and a lot of smiles as a woman wearing a feather headdress and high heels entered the ring carrying a flaming torch. She was tall with a small waist and a full bosom. She smiled back at the crowd, holding the torch aloft. She turned slowly in a circle to a resounding applause embellished with catcalls. She took deep breaths of appreciation, heaving her cleavage.

Nodding to Rolando, she lowered the torch until it touched a silver ring. It immediately burst into flame. The lone cat roared so loud that the crowd fell silent for a moment and Rolando paused to look at the lion before turning his attention back to the tribe.

He cracked his whip and the lions obediently hopped down from their perches and formed a single line. Each lion took a turn jumping through the ring of fire before returning to her perch. Each brave leap was rewarded with applause and smiles.

I couldn't clap if I wanted to. I attempted to free a napkin from under the bag of peanuts. The bag of popcorn wobbled precariously on my leg, tipping dangerously as a few kernels fell to the floor. I relaxed my movements and took a bite of the hot dog.

"And now, ladies and gentlemen" came the amplified voice, "for his next act, Rolando has requested complete silence. Please refrain from making any loud noises or any sudden movements. Rolando will now brave the mouth if the savage beast." The crowd quieted and concentrated on the center ring.

Rolando walked behind the lone male cat and thumped him on the butt with the handle of his whip. With lightening speed the lion turned, striking out with his paw, missing Rolando's face by inches but managing to connect with his shoulder, tearing his jacket and throwing Rolando off balance.

Rolando's face flushed as he stumbled then regained his footing. He said something to the cat, but the audience couldn't hear it. The lion answered Rolando by extending his neck menacingly and furrowing his brow. A growl that came from deep within his chest rumbled as he bared his teeth.

Appearing to recover from the incident, Rolando attempted to assure the uneasy crowd that everything was okay. He removed his hat and held it aloft, bowing to the crowd as if this was a part of the act. But, it wasn't.

"Benny, are you watching this?" Leo asked, unable to tear his eyes away from the center ring. "This is great entertainment."

I watched helplessly as the bag of popcorn began to fall. I tried grabbing for it, but my attempt only knocked the bag to the floor sooner and jarred my shoe loose from my toes. My shoe fell under the bench in front. With my hand still clutching a half-eaten hot dog, I leaned over to retrieve it.

The cat watched Rolando as he turned in a circle receiving applause. He hopped down and paused as if to get his bearings, then moved cautiously toward Rolando, snapping his tail. The lion was quiet now as he lowered his torso to the ground and crept—stalking. From the corner of his eye Rolando caught the movement of a blonde mane and turned quickly, dropping his hat. The lion's eyes were fixed, unwavering, loathsome. Rolando watched the cat wide-eyed as he backed away cautiously. His face no longer flushed, Rolando had turned pale and his body shook.

"Benny," said Leo, tearing his attention away from the center ring just long enough to see Benny duck under the bench. "What are you doing? You're missing the best part."

Without warning the cat sprung . . . my hand groped blindly for my shoe; it was dark under the benches . . . the lion's paws were quick and strong, striking their target unmercifully, tearing the left side of Rolando's face away and knocking him to the ground . . . finally, my fingers found a shoe lace as the hot dog slid out of its bun and fell to the floor at my knees . . . the cat was quick and thorough—before Rolando had time to realize what was happening, the lion took the top half of Rolando's head in his mouth and chomped his jaw shut . . . I heard gasps from the audience and was disappointed I was missing the 'best part', but my shoe had to be retrieved and my sock felt damp from the mustard I was stepping in . . . the lion's incisors, each nearly four inches in length, pierced Rolando's temples and sank into his brain.

"Jesus Christ!" I heard my grandfather mutter as Rolando's body convulsed and shook uncontrollably. . .

. . .The vehicle bounced, rocking gently, jarring me from reminiscing. I sat in the back seat behind my mother watching my father drive. The oversized tires of the rumbling Buick rolled swiftly over the asphalt; the journey to Irma and Harold's would take a full two hours.

Pedal pushers hugged my mother's legs. She stretched, pressing the balls of her feet into the floorboard. The air that slipped into the car through the window vent tossed her short curls about like tiny kites caught in the wind.

"Will everybody be at the picnic?" I asked, leaning forward and holding onto the back of the front seat. "I can't wait until we get there."

"I doubt if everyone will be there," my father answered softly. He looked sidelong at his wife and their eyes met.

"Will Grandpa Worthy be there?"

"We'll just have wait and see." She watched a train in the distance running parallel to the car. Sometimes they seemed to keep pace, then a moment later the Buick gave the illusion of winning the race. Eventually, the locomotive pulled away, its eighty cars disappearing with it and she was overwhelmed by an emptiness, as if the past was pulling away from her, too.

"What time did you tell Irma we'd be arriving?" My father asked.

She looked over at him. They were on their way to her sister Irma's for a family picnic. It was Sunday. The Fourth of July. Pete was just a bad memory, an ex- husband she wished she'd never married. The only thing good that came out of it was Jeffery.

Now she had Hank, an attorney fresh out of law school and doing well. Her five- year-old son, Benny, was sitting behind her in the back seat thumbing through a comic book. She had a good life, now. She was determined to be happy.

Two pies: a Washington Nut pie, otherwise known as a pecan pie, was covered with a dishtowel, and a lemon meringue pie sat in a white box on the front seat next to her. She had made each from scratch and was very proud of them. It had taken her weeks to finally perfect the crusts, but now she had it down.

She held her new daughter, Holly, in her lap. She'd never known a child to sleep so much. Such a strange girl. Neither Jeffery nor Benny ever slept that much. She hadn't seen her son, Jeffery, in quite some time . Her heart ached.

"Hello. Is there anybody home?" It was my dad.

"What?" She said shaking off the past.

"I said, what time did you tell Irma we'd be arriving?"

"I didn't exactly tell Irma that we *would* be arriving."

My dad looked at her questioningly as a sly grin graced her lips.

He downshifted for a curve. "I don't get it. Is there something you're not telling me?"

"Irma," my mother said, crossing her arms, "has no idea we're coming."

"Why's that?" he asked, afraid of what her answer might be.

"I told her you weren't feeling well and that we wouldn't be able to make it."

"That I wasn't feeling well?" he repeated, surprised.

"But, you're feeling okay now, aren't ya Dad?" I asked, sitting forward.

"Oh, a feel great," he said, grinning, amused and shaking his head knowingly.

"Boy," I said, "She sure gonna be surprised when she sees us, huh?"

"You don't know the half of it,'" he replied,. "Light one of the for me too, will you? I feel a crises coming on."

My mother gazed back out the window. The telephone wires dipped from pole to pole like slack mooring ropes. Giant billboards were quick reads and momentary escapes from the monotony of travel.

"I'm tired of all the bullshit," she said, handing her husband a lit cigarette. "They've always arranged it so Jeffery and I would never be in the same place at the same time." She took a long drag and hissed a plume of smoke into the car's interior. "I'm sure he's noticed it, too.

They must think we're both morons." I watched my mother sit back in her seat, determined. "But, not today, by God. Not today."

My father laughed. "Boy, are they gonna shit."

I wasn't exactly sure what my parents were talking about. But, I knew who Jeffery was, and it wasn't hard to figure out that something very out of the ordinary was about to happen.

I don't know why I suddenly thought of it, but I remembered another encounter that happened when I was much younger.

One morning, when I was three and a half, I meandered down the sidewalk to a neighbor's. Mrs. Graysome was a friend of my mother's. Four months earlier she had given birth to a fat blonde boy who cried more than he laughed and took frequent naps to withdrawal from a world he found too overwhelming.

I was always made welcome in the Graysome home, so in my daily excursions it wasn't unusual for me to wander into their dwelling. On this occasion I found Mrs. Graysome lying on a couch nursing her son. The couch seemed much too small for her Rubenesque figure and I marveled that the tiny wooden legs could support so much weight. The couch seemed to sag in the middle and rise up at both ends.

"You're up early this morning," Mrs. Graysome said to me as she watched me shuffle into the room. I smiled then watched Mrs. Graysome nurse while keeping my hands in my pockets. Mrs. Graysome's breasts were large to begin with, but when her milk came in they grew to an enormous size. She could have easily smothered her child if she wasn't careful.

"Got any chocolate milk in there?" I asked, nodding to Mrs. Graysome's breasts. I was hoping the answer would be "yes"; I sure did like chocolate milk and wondered what it would be like to suck on a big titty.

"Nope, just regular white milk," Mrs. Graysome answered, suppressing her amusement.

I watched for a while, my hands still buried deep in my pockets. "So, how much you got in there," I asked, marveling at Mrs. Graysome's breasts, "about fifty gallons?"

Mrs. Graysome laughed lightly. "Not quite that much, I'm afraid. But, quite a lot, I'm sure." The answer seemed to satisfy me as I continued to watch her nurse. I had been bottle-fed, so was curiously attracted to Mrs. Graysome's big white breasts.

"What have you got in your pockets, Benny?"

I smiled sheepishly and moved my fingers inside my pockets. "Friends," I answered. "They're sleeping."

"Who's sleeping?" Mrs. Graysome asked.

I pulled a hand from my pocket. I kept my fist loose so as not to damage its contents.

"What have you got there?"

I turned my hand, palm-side up, and slowly unfold my fingers. "Friends" I answered again.

Mrs. Graysome's eyes widened as she rose up on one elbow to get a better look. As she rose her nipple slipped out of her son's suckling mouth with an audible pop just before she gasped.

I ran my finger over the inert bodies of the three dead mice that lay in my palm. "They're brown," I said to Mrs. Graysome just in case she wasn't up on her colors. "We have to be very quiet," I whispered, "or we might wake them."

"Where," she gulped, "did you get them?"

"On a garbage can in the alley," I answered. "They were laying right on top—sleeping."

"Does your mother know about your little . . . friends?" Mrs. Graysome asked, aghast. The mice were small, but perfect in every detail. Their coats were full and thick, their tails, long and skinny. Their eyes were closed and the lifeless animals did indeed appear to be sleeping.

"Nope. Not yet." I continued to pet the mice with my finger as I spoke. "They're so soft," I said.

Mrs. Graysome's cat, a five-year-old short-tailed tabby, hurried into the room. He'd been drawn by the exceedingly agreeable aroma coming from my extended hand.

When he opened his mouth to speak his throat was so full of drool he could only croak. He shook his head and swallowed hard, but the secretions only flew about the room or spilled over his lips and dribbled onto the floor.

Mrs. Graysome felt something wet hit the corner of her mouth. "Get out of here, Bentley," she snapped.

Bentley acknowledged his name with a quick glance, then turned back to the delightful tidbits sitting in my hand.

The cat had a few fears, but the bark of Mrs. Graysome's voice wasn't one of them. His tail had a crook in it from being run over one day by Mr. Graysome while backing their pickup out of the garage. A dog got ahold of Bentley when he was barely six weeks old and shook him so severely his ears became disconnected—now, one ear always stood up straight, while the other one flopped, covering his bad eye. Before he was two months old, while playing too rough with another in the litter, his eyeball was scratched so deep it soon clouded over and he never saw out of it again.

"Scat!" Mrs. Graysome said harshly.

Bentley didn't hear or ignored her altogether. He placed his front paws on my shorts and buried his nose into the fabric of my pocket, inhaling deeply and rolling his eyes back in his head, relishing the savory delectables.

"I've got more," I announced, ignoring Bentley and reaching into my other pocket.

"No. No. Don't show me," Mrs. Graysome said frantically, sitting up. "You go home and show your mother."

Showing my little friends to my mother sounded like a pretty good idea, but Mrs. Graysome hadn't bothered to cover herself when she sat up. One of her breasts had spilled over the top of her sleeping gown, undulating back and forth.

The delicious scent of dead mice was so enticing to Bentley, he became aroused.

His tiny red penis slide out of its sheath while he pawed rhythmically at my pant leg.

As Bentley grew more excited, his claws extended further, tearing through the fabric of my pants and breaking the skin beneath. I jumped back suddenly.

"Stop it," Mrs. Graysome barked, leaning forward and smacking Bentley on the head. Bentley laid his one moveable ear back, cowering, but stood his ground. "Go on," she said, lumbering onto the floor on all fours, trying to push Bentley away. "Get out of here."

My attention drifted away from my "little friends" to Mrs. Graysome who was scooting around on the floor. Her other breast had somehow managed to tumbled out of her sleeping gown, too. Now, both pendulous breasts wobbled back and forth as she attempted to subdue Bentley. I stared in wonderment as her nipples brushed the floor, leaving little trails in the dust.

So complete had my attention been diverted; I didn't notice the mouse spilling out of my hand. Bentley, on the other hand, saw the mouse while it were still airborne and moved with such agility that it landed in his mouth before it hit the floor.

In her effort to protect me, Mrs. Graysome shoved me out of the way. I teetered for a moment then fell backwards, breaking the fall with my elbows. Mrs. Graysome took little notice, as she grabbed Bentley by the throat and tried to pry his mouth open.

"Give it up you son of a bitch," she muttered through clenched teeth. Bentley's jaw had become a vise. So determined was he to consume the dead rodent, he chose to swallow it instead of giving it up. But, the tasty morsel was too large and became lodged as Bentley gagged. His hindquarters stiffened and he began jerking about on the floor, choking and convulsing.

"Open your mouth you stupid asshole," Mrs. Graysome said angrily as she pinched his lower jaw, forcing it to release. Using all her strength, she reached inside while Bentley continued to choke.

Fearing for the life of one of my dear "friends," I scrambled to my feet and stomped on Bentley's tail in an effort to help Mrs. Graysome. If Bentley could have shrieked he would have, but his voice box would only allow a pathetic gurgle to escape.

What with Bentley growling and thumping around on the floor and Mrs. Graysome cursing and shouting while trying to subdue him, I figured now was as good a time as any to head back home.

Besides, the thought of one of my "friends " being eaten was more than I cared to think about.

Just for good measure, I gave Bentley's tail another good stomping before gathering what was left of my "little friends" and leaving by the back door. "Bad kitty," I muttered as I maneuvered down the stairs. . .

. . .The oversized tires of the rumbling Buick rolled swiftly over the asphalt; the journey to Irma and Harold's would take another hour or more.

I thumbed though a comic book for a while, but was soon bored. Yesterday had been unusually eventful and exciting. It had started out like any other summer's day, I suppose.

Early in the morning, while my mother still slept, I made my way to the gravel alley behind our house to catch bumblebees in a jar.

The hollyhocks grew so tall they towered over the roofs of the garages. The flowers that unfolded were two-hands wide. The bumblebees that flew in and out were so lipid they had to take running starts to lift off. More than once an overweight bee failed to reach enough speed and tumbled helplessly over the petals, falling to the ground.

Catching bumblebees crawling on the ground, their wings buzzing frantically, was not my idea of fair play. The enjoyment of catching a bumblebee was the risk of being stung. Bumblebees so fat they couldn't get off the ground were more sad to watch than fun.

I knelt and placed a twig in the path of the struggling insect. It climbed aboard willingly, then froze as I lifted the twig above my head.

"What are you afraid if—heights? Come on, little guy," I said, jiggling the twig. "Fly." The bumblebee buzzed its wings, lost its balance, and fell to the ground again. Bees didn't seem to be as aggressive as they used to be. Like last summer at Aunt Irma's.

Aunt Irma was just one of seven aunts still alive. My mother had seven sisters and one brother. All of my aunts called their brother,

"Brother." I called him: Uncle Brother. It wasn't until I was years older that I realized "Uncle Brother" had a Christian name as well.

Most of my mother's sisters married farmers. And, the fields they tilled all touched one another. Throughout the year, on major holidays, or when the Grandparents celebrated an anniversary, and especially during the three months of summer, large family get-togethers were held, congregating under one roof—usually Aunt Irma's.

Irma and Harold's farmhouse was the grandest of the lot, set off from the main road, high on a hill. Sloping hills of blue grass surrounded the house and though a dandelion was seldom found, wild buttercups dotted the grass.

One morning when the dew lay like a wet sheet over the lawn, and I wasn't mindful of my step, I slipped and slid down the hill on my back, my journey ending abruptly at the dirt road below.

When I sat up and looked back I was amazed at the considerable distance I'd traveled. The seat of my pants and the back of my shirt were soaked through—cold and damp in the brisk morning air.

"I can think of better ways of taking a bath."

I looked up at the lanky figure standing over me. What little sun made it through the thicket of pine was in my face and blinding. The only thing I knew for sure was the voice had come from a girl.

"I usually sit on a cardboard box to do that," she said. "You don't get quite as wet that way."

"Can you keep a secret?" I nodded my head "yes."

"Once," she began, "I got up before Aunt Irma did and ran outside, took off my pajama bottoms and slid down the hill on my bare ass." She laughed and slapped her thigh in amusement. "I tell ya," she said, "I'm a real card, I am.

"I'd like to do that."

"You would, huh?" I nodded.

"Okay, tomorrow morning, then. I'll wake you up early and we'll go do it," she promised.

"Really?"

"Sure as shootin', buster." She set down her pail and a dozen earth worms squirmed over the side. She stuffed them back in then placed her bare foot over the opening to prevent further escapes.

"Do you know Aunt Irma?" I asked, beginning to like this strange girl.

"'Course I do. She's my Aunt, too. How old are you?" she asked, looping her thumb under the strap of her bib overalls.

"Four," I answered.

"Four?! Well, poop my pants and call me Dumpy."

I laughed at her words and she smiled back. "Well," she said, "I'm six, nearly seven. That means between you and me, I'm the adult and you have to take orders from me, 'cause I'm the boss. Okay?"

"Okay." Made sense to me.

She was standing with one hand on her hip. Her other hand was balancing a fishing pole on her shoulder.

"You going fishing?" I asked, getting to my feet and brushing off my pants.

"Looks that way, don't it?" Her red hair was pulled back in a ponytail. She smiled with her hazel eyes and shook her head. "You must be a city boy."

"I guess I must be," I admitted.

"I'd ask if you want to go fishing with me, but by the looks of ya I'm afraid you might scare the fish off."

"That's okay," I said, "I don't know how to fish anyway."

"Maybe it's about time you learned, then," she said, walking closer. "What's your name?"

"Benny. I'm just visiting."

"Who ya visiting?"

I pointed back at the farmhouse. "My Aunt Irma."

"Is that right?" she said, smiling. "Since that's my Aunt Irma too, we're probably related."

"Think so?" I was surprised and pleased.

"Is your mother's name, Joanie?" I nodded in the affirmative. "I've heard of you, but never thought I'd ever meet ya. Your mother is my mother's sister, Doris."

"Really?"

"That makes us cousins." She extended her hand and we shook. "I'm Nellie. Nellie Fairfax. We own the dairy farm down road."

"You know how to milk cows and all?"

"Sure as shit do," she said. "So, you want to go fishing? We could make another pole out of a tree branch and I've got plenty of worms." She looked down at the tin can and wiggled her toes.

"I'll have to ask to see if it's okay first. I don't think anybody knows where I am."

We trudged up the wet slope in the rear of the farmhouse, grabbing hold of clumps of grass to aid our ascent. When we reached the top, we stopped and caught our breath. Before us, the farmhouse loomed in the mid-morning sun.

Seven white columns supported the shingled porch awning that wrapped around the house. Uncle Harold built the porch at Irma's request, displaying a fine talent for carpentry and planning. Over-sized hanging planters brimming with fuchsias of purple and pink surrounded each column.

Two porch swings, one on the side of the house, near the living room windows and another in the rear, facing the woods, rocked and swayed—pushed by the summer winds.

"You the only one visiting?" Nellie asked.

"Heck no, everybody's here, or will be pretty soon. Today's the family picnic," I announced.

"Yeah, I know. This is the first family picnic you've been to out here. I guess that means Jeffery won't be here."

Nellie was right. Jeffery wouldn't be there. It was an unspoken truth. If my mom or any of our family were about, Jeffery wouldn't be. I knew who Jeffery was, but had never seen him.

We found Aunt Irma in the kitchen standing at a counter whipping cake batter with a long wooden spoon. The scent of powdered sugar and vanilla extract floated in the air. When Irma figured the batter was nearly whipped, she rinsed her hands and called one of the cousins to take over.

"Stevie?" Steve looked up from the cake he was frosting, a smudge of chocolate icing in both corners of his mouth. "Looks like you got more frosting on yourself than the cake. Come on over here, honey." Steve reluctantly returned the spatula to the frosting bowl. "You can take over whipping this batter. Whip it till it's smooth and creamy. When it looks and feels like a thick milk shake, let me know. All right?" Steve nodded and set to work.

There were half s dozen cousins in the kitchen that morning, all busy working at their assigned tasks. And, though another grownup might have found the noise and occasional bickering to be too much, Aunt Irma never seemed to mind.

"I'm getting low on eggs," she shouted above the din. "Who wants to go to the chicken house and get me a dozen?" Several hands went up. In cousin Theresa's hurry to volunteer, she knocked a bottle of heavy cream from the table. It crashed to the floor, scattering cream and broken glass the length of the room. The children quieted and stopped their work.

Irma possessed a storehouse of patience. It never ran out. Even when little Joey peed his pants one day while standing in the kitchen watching Aunt Irma spread marshmallow frosting over a three-layer cake.

I was sitting on the edge of the counter licking a beater clean when I watched Joey's pants turn dark. The puddle formed a small lake around Joey's shoes.

Irma put her spatula down, walked casually to the sink and rinsed her hands. The water that poured from the faucets was always lukewarm and bubbly.

Tears were streaming from Joey's eyes. "It's all right, honey," Irma said, drying her hands. "It's certainly nothing to cry about. Accidents just happen sometimes." She knelt down and gently took Joey's hand. "You didn't do it on purpose, did you?" Joey shook his head. "Well then, there's nothing to get upset about then, is there? Come on. I'll run some warm water in the tub and you can take a bath while Aunt Irma throws your trousers in the washer. They were dirty anyway."

It was always like that with Irma, even on the days of the big family picnics. Relatives would travel from all over, some driving all night before reaching Irma's farm. When they arrived they were encouraged to eat. There was always something on the stove, and since it was a pot luck, the tables would be brimming with many a tasty dish.

Of course, Irma would have several cakes already baked and set aside ready for frosting. When she figured enough guests had arrived she'd wonder aloud if any of the young cousins would mind giving her a hand in the kitchen.

There was usually a scramble for the kitchen. If Irma found there were too many children for the number of chores, she'd begin looking for new chores. This gave the parents some peace away from the kids and Aunt Irma a chance to be near the children.

"Now, don't go fretting over spilled milk, Theresa, hon. You just get Philip there to help you clean it up. Be careful of the glass. Stevie, you give Philip a hand."

When she turned, she saw me standing in the doorway with Nellie. "I expect you've been off contemplating the universe somewhere," she said to me. "Nellie, you look you're about to go fishing."

"Yep."

"I suppose you want to go with her," she said to me. I nodded. "Well, I don't see anything wrong with that, but I think you're a little young to go by yourself."

"I'm not too young," I protested.

"I know, honey," Irma agreed, "But, if anything happened to you, your mother would hang me out to dry. Paul Lee!" She looked around the kitchen.

"Paul's out in back of the barn," said Theresa.

"Benny, you and Nellie go to the chicken house and get me a dozen eggs, then tell Paul I want him to go with you two to the fishing pond."

Outside, the air was still brisk in the soft morning sun; the aroma of manure drifted over from neighboring farms. Robins dotted the sloping lawns. The grass was still wet with morning dew and the blades stretched themselves stiff to drink in the moisture.

Nellie and I stopped to watch the birds feed. Their long legs skittered across the lawn in sudden spurts. One bird had ahold of a worm that was reluctant to leave its home. The bird tugged, stretching his prey like elastic until it snapped, part of it sliding down the robin's gullet, the rest disappearing back into the wet earth.

Inside the chicken house, the floor was a sea of yellow. Many hundreds of baby chicks scrambled for cover, running aimlessly, chirping.

Irma and Harold hadn't always been interested in raising chickens. It was quite probably the furthest thing from their minds. But, one unusually warm day, on a shopping trip to the district of Hillyard, Irma left several cartons of freshly bought eggs on the floor of the stepvan while she and I ran into the hardware store to buy supplies.

"Morning, Bob," said Irma, letting me enter the store first and closing the door behind her.

"Well, hello there, Irma. Who's that you've got with you?" Bob was wearing a blue denim bib-apron with a yellow pencil parked behind his ear.

"This is my sister Joanie's little boy—Benny. He's spending a couple of weeks with us this summer. Benny, this is Bob."

Bob stooped over and shook my hand. "Pleased to meet ya, young man," he said. He had a warm smile and kept his eyeglasses perched on top of his head where he was rapidly going bald.

Bob's Hardware Store had wooden floors, a high ceiling and incandescent lighting. Bob got rid of the fluorescent lights soon after he bought the store from Tim. Back then it was known as Tim's Hardware Store. Bob bought Tim's Hardware Store a short time after Tim suffered a stroke.

It wasn't one big stroke that Tim had, but hundreds of tiny ones. He wasn't rushed off to the hospital in the middle of the night with sirens screaming, nor did the little strokes incapacitate the muscles on either side of his face or make him a cripple.

The transient ischemic attacks visited him while he slept. He awoke the following morning appearing much as he always did, but not entirely sure where he was.

Tim suffered one terrible symptom as a result of the extracranial interruptions to his arterial blood flow; he began to experience a problem with his memory. He had a tough time remembering anything.

Likewise, his wife was afflicted by a condition that rendered her involuntarily bulimic. She seemed fine; her mind could remember instances in her childhood in minute detail; she could recite passages from the Bible without a flaw; she knew Robert Service's poetry by heart and chanted versus while rocking in her chair by the open window when she was alone. She knew the time and date of every relative's anniversary, birthday and death.

But, she couldn't remember the last time she brushed her teeth and would end up brushing them five or six times in a day. That was harmless enough, but when she couldn't remember the last time she ate, she would simply eat again—and again—and again until her stomach could hold no more and she found herself in the bathroom throwing up. Since she didn't know she couldn't remember the last time she ate, she assumed she had a case of the stomach flu.

Tim had no idea why his wife looked so ill, but if Bonnie said she had the stomach flu, then the stomach flu was what she had. He knew better than to argue with her.

When he volunteered to make dinner, she ate well enough, but soon afterward s he would hear Bonnie in the bathroom throwing up. The food must have been all right, after all it settled just fine in his stomach. Bonnie definitely had the flu. Strange that the flu would hang on for so many months though.

One day, after a perfectly comfortable evening visited by numerous ishimic attacks, Tim awoke feeling unusually refreshed and went to work as usual.

At the end of the day, he closed up shop, stepped outside, turned the key in the lock, and with a paper bag containing the week's earnings grasped firmly in one hand, and started off for home. He hadn't traveled far before he realized he couldn't remember where he lived.

He stopped and looked back. The street traffic was light, but then, this was Hillyard and traffic was always light. A small dog recognized him and laid his ears back to let Tim pat his head.

"Shit," Tim muttered straightening up, "where the hell am I?" He knew something should look familiar, but dammit, nothing did. "I must live around here somewhere." Across the street two children waved and Tim waved back. The light changed and since Tim was standing at the crosswalk, he figured he might as well cross with the light. "Oh well," he shrugged, "if I just start walking and pretend like I know where I'm going, I'm bound to recognize something sooner or later."

Unfortunately, after walking for more than two hours, nothing looked even remotely familiar. Any direction Tim may have chosen would have taken him farther away from his home. For the last eighteen years, Tim lived with his wife in an upstairs apartment directly over the hardware store he'd just left.

Tim ended up at a bus stop where he sat down on a bench to rest. He carefully placed the bag next to him and looked at his hands. He marveled with dismay at the texture of his skin and the number of wrinkles that rippled over its surface. The hair on his hands was white now and the skin on his knuckles was loose and wobbled about when he nudged it with a finger.

"I must be getting old," he thought. "How old am I, anyway? Let's see, I was born during the last century, sometime. Christmas! No. Christmas Eve! Yeah, that's it—now, we're getting somewhere." Tim pondered for a moment and scratched the dandruff from his eyebrows. He watched the dry skin fall before his eyes. When he was a little boy he would watch the snow fall from his bedroom window. The shutters had been thrown wide and the moon shown on the new fallen snow. The air was so cold it froze the mucus in his nostrils and made the muscles of his cheeks stiff.

"Timmy! You close those shutters and get back under the covers. That cold air will make you sick. Come on, now." His mother swooped him up in her arms and laid him gently on his bed. The mattress was thick and lumpy, but soft and comforting. His mother had sewn the quilt that she tucked into the sides of the bed. The sheets were pure white and stiff from just being washed.

"Did you say your prayers?" Tim nodded. "If you didn't say your prayers and you're saying you did, God will know. He always knows."

Tim smiled. There was no one in the world he admired more than his mother. "Tell me a story," he demanded gently.

"A story? I don't know any stories."

"Please," Tim pleaded, touching his mother's hand.

She smiled in spite of herself and lowered her weight onto the bed. "I don't know any stories. All I know is the truth."

"Tell me the truth then."

"All right." Mrs. O'Sullivan cleared her throat and began.

Tim laid back and sank into the pillow. He watched his mother's lips move as she spoke. Her words were spoken softly and slowly as she spun web after web of intrigue and imagery. Her tales of adventure took Tim down through the ages. She spoke of myths and gods and rulers and kings. As his mother lulled Tim to sleep with her words, his eyes left her face to wander the room; he listened to the wind as it slipped past the shutters and imagined he heard the snow flakes as they landed on the window sill, outside.

Like his eyes, his mind wandered too, and Tim remember last year, before his father died. It had been a night much like tonight. The snow had fallen for hours. Tim placed a pillow on the floor at the base of his bedroom window and pushed the shutters open. He sat on his knees, the pillow softening the load, and watched the horse-drawn carriages and cabs make their way up the avenue. Sleigh bells rang in the night and voices were muffled by the descending curtain of snow.

Down below, he watched the comings and goings of Boston. Everywhere he looked his six-year-old mind grappled with the myriad of life's textures—the moaning of the ships' horns bellowing in the winter fog, the "stiltman" pegging along the sidewalk directly below his window, lighting the lamps.

Soon, he saw his father walking home after work. He was a couple of blocks away, a long wool scarf wrapped tightly around his neck, stepping carefully over recently plowed mounds of snow. Every now and then he stopped and coughed violently into a handkerchief, then recovering, his head spinning from the sudden paroxysms, he carefully refolded the rag and tucked it back into a pocket.

Tim could see the books his father carried, clutched tightly under one arm, and worried he might drop them each time he had a coughing attack. But, he never did. Books were treasures and his father guarded them as though they were living entities. As Head Librarian, Mr. Lawrence O'Sullivan had his pick of the finest new volumes that arrived almost daily through the mails. And despite this recent onslaught of the flu, nothing could stop Mr. O'Sullivan from sharing these newest treasures with his son.

"Wait until you see what I've home for us to look at tonight, Timmy," he announced, entering Tim's bedroom. Flakes of snow still clung to his hair and hadn't melted yet. He laid the books at the foot of Tim's bed joined his son on the floor at the window. He rested his knees on the same pillow and looked out at the night. "What are we watching?" He asked, then coughed.

Tim shrugged and smiled. He wasn't watching anything in particular, now. But earlier in the evening he'd seen things he thought were quite amusing.

Earlier, Tim watched Old Man Worthy close up his mercantile store for the night. The hem of his long coat brushed the snow on the sidewalk as he turned away from the street and pushed his key into the lock. His cab awaited, the black horse harnessed to the carriage breathing evenly in the cold air, steam puffing from his nostrils.

Then, magically, as if guided by an invisible hand, but really because it had been installed wrong side up, the bolt securing the carriage to the harness dropped into the snow. The horse sensed a release in tension and took a step forward.

Surprised, he twisted his head to look. He was definitely farther from the coach than he had been, and as if test this, he took another step and watched the distance grow. He swished his tail and considered the possibilities. The snow fell lightly, bloated flakes falling quiet, swirling past the gaslight.

Before his brain could make much sense of it all, the front end of the cab fell nose first onto the street. Cushioned by a pillow of snow, it made no sound and Mr. Worthy noticed nothing.

Taking one last fleeting glance at Mr. Worthy, his back still turned, stomping the snow from his boots and taking the time to light his pipe, the horse trotted quietly away.

Tim covered his mouth to stifle his laughter as he watched with amusement from his elevated vantage. Soon the horse was two blocks away, with no thought of a destination, trotting up the middle of the avenue, his head held high.

At times, the horse looked behind, worried someone might be close behind—waving his walking stick, his coat tails flapping in the wind, his cheeks ruddy and round, puffing and screaming—in hot pursuit. It scared him to think that this same walking stick might land at any minute on his rear to deliver a well-deserved wallop. The fear of such an onslaught only encouraged him trot faster, increasing the distance between him and Old Man Worthy.

Mr. Worthy, for his part, was lost in his own thoughts. Though he occasionally stomped his boots on the sidewalk, inhaled the rich smoke from his pipe and looked down the length of his coat to satisfy himself that each button was in its proper place, he procrastinated.

He considered stopping for ale first, or perhaps grabbing a bite to eat at McRory's before going home. His wife, Dorothy, would be asleep and the house would be as quiet as death, except for the agonizing ticking of the grandfather clock in the foyer. If he had to do it over, he would never have bought that damned clock—the pendulum swung too far and too slow, taking far too long between ticks.

He pulled a watch from his pocket and unsnapped the cover. 8:15. He brought the watch close to his ear and listened to the rapid ticking. That was more like it. He smiled with satisfaction and replaced the watch.

"Damn, I love the snow," he said aloud and stomped his boots again. Old Man Worthy strolled to the corner, just a few feet away, leaving the protection of the wood awning and looked up at the night sky. Snow fell from the heavens and landed on his face. He stuck out his tongue and caught the cold flakes as they landed.

"Don't you just love the snow Beauty?" he asked, turning to his carriage.

At first Mr. Worthy wasn't sure what the hell he was looking at. The carriage sat at an odd angle, its rear raised to the sky, the cross-ties nosed into a mound of snow. His horse was nowhere in sight. He looked up the street, but Beauty was gone. He shook his head and almost laughed, but the situation was far too serious to be made light of.

That was the stupidest horse he ever owned. If God ever made a more stupid horse it wouldn't have lived beyond its first year. Dorothy was going to be so upset. First Arthur, their son, running off, traipsing out west to seek his fortune, a kid not even thirteen years old thinking he knew it all. And now this.

Mr. Worthy sat down on the curb and gazed at his useless carriage. He'd have to hail a cab sooner or later, he figured. Jesus, that horse was dumb! Where the hell did he think his next meal was going to come from? It wasn't as though he could stop and graze anytime he wanted to. It had been snowing for a week and was bound to snow for another before it let up. By that time, Beauty will have starved to death and ended up a carcass somewhere for stray dogs to feed on.

Arthur told him before he left that Beauty wasn't very bright. "She won't stray Dad. She's loyal even if she is stupid." Beauty had been Arthur's horse. If anyone should know, Arthur would. Mr. Worthy hoped his son was right. He could almost hear his son's voice when he remembered his words. He missed his son. Arthur was an awfully smart kid, he thought.

When Arthur was three he was already able to read. By age four he convinced his father that lying about his age was no great sin and that going to school a year or so early would be good for him. He had been right. No one was ever the wiser and Arthur excelled in every subject. Physically, he was always a little smaller than his peers, but he didn't seem to mind.

When he was twelve, Arthur graduated from the eighth grade and instead of choosing to go on to college to study medicine or law, he announced one evening he wanted to be a cowboy and headed out west. Sometime later Mr. Worthy learned that his son had worked on dairy farms in the Midwest for a spell, then met a girl one day and fell in love. Myrtle was her name. They married, then traveled farther out west and settled in a little town in the state of Washington—Steptoe.

James Worthy sat on the curb in the falling snow, shaking his head, remembering the shock of Arthur's announcement that evening at dinner. A cowboy, James thought, looking at his cock-eyed carriage dipped nose-first into the snow bank.

Arthur had even written him a letter. Mr. Worthy reached into his pocket and removed a worn paper envelope that was beginning to tear at the edges. The letter had been read and reread so many times the folds were beginning to split. Arthur wrote a good letter. He had a fine sense of humor and the older he got the wiser he became.

Little Timmy O'Sullivan watched from his second floor window as Old Man Worthy unfolded the letter and began to read.

January 26, 1897

Dear Mother and Father,

It's been snowing here for nearly a week, now. At times I wonder if it will ever stop or if the snow will just keep on falling forever and we'll have to buy ourselves snowshoes and learn to ski.

Our neighbors, Clyde and Mary Kellogg, let the snow pile up in the back of their wagon. The snow just kept falling and the back of their wagon just kept getting fuller. I said to him "Clyde, if I were you I'd get that snow out of there before it got any deeper."

Well, he just poo-pooed me and late yesterday afternoon, I happened to be outside when Clyde decided to hitch up the wagon and head on into town.

I was standing on the front porch at the time, amused that Clyde would even attempt such a thing. The wagon must have held nearly a ton of snow. Well you can guess what happened.

I guess congratulations are in order. Looks like you're grandparents now! That's right, Myrtle gave birth to a beautiful baby girl just today.

We've decided to name her Irma. And, I'm sure she'll grow up to be a real fine lady someday.

Sure do miss you guys.

Love,
Arthur & Family

Sure wish Arthur would have stuck around, Mr. Worthy thought. He ran his finger over the fibers of the paper, the feeling somehow drawing him closer to his absent son. He refolded the letter and slipped it back into its envelope while gazing absently at his carriage, which looked more like a teeter-totter than a mode of transportation, now.

Just then, James looked up to the sound of a cab passing by. "Hey, there!" he hailed, raising his hand. Moments later, comfortably settled within the confines of the cab, James once again pulled Arthur's letter from his pocket and began reading it again.

Little Timmy watched the cab pull away. It grew smaller as it headed up the avenue and finally disappeared around a bend in the road. From his second-story vantage, Boston was a storybook model of brick dwellings, cobblestone roads covered with packed snow and chimneys belching gray smoke into the cold winter air.

His father was settled beside him of his knees sharing the same pillow, looking out into the night. "So, what are we watching?" he coughed. Timothy shrugged. Nothing. Everything. He wasn't sure.

"You're watching the reflection of the sun bouncing off the moon and washing over the surface of the night's snow," Tim's father said. "You're just being and enjoying being. Nothing wrong with that." The family cat climbed onto the pillow and joined them. He settled onto his side and licked his coat while purring.

"Do cats believe in God," Tim asked, wondering aloud.

"They don't have to believe," his father replied, then coughed. "Faith isn't required of animals—they accept their fate as a matter of fact. Only man questions his existence."

"I'm not sure if I believe in God," Tim confessed with some confusion and uncertainty.

"I'm sure God will be very disappointed to learn that," said Lawrence, coughing again.

"Will He be mad at me?" Tim asked, a little worried.

"He understands," Tim's father replied, smiling at his son's contradictions. "Not believing in Him won't change anything. He won't become any less powerful or any less merciful and He'll still love you just as much as He always has. There'll always be a place for you near Him anytime you want to take it. You don't have to believe in God just because your mother and father do. It's probably better that your faith come from inside anyhow. God will be delighted when He learns you found Him on your own."

Lawrence always spoke to his son like an adult. There was never any baby talk or dumbing-down of vocabulary.

Tim's father was extremely ill the next day, but went to work anyway. His pride and work ethics prevented him from remaining home. It was inconceivable to him that the library could function efficiently without his ever-watchful diligence. He refused to listen to his body and followed the presets of his mind. So, the next day, instead of resting, he took the virus to work with him and promptly infected his staff.

That year, the flu epidemic that attacked three-quarters of the world claimed nearly two million lives. Tim's father was one of them. Four weeks after coming down with fever, Lawrence O'Sullivan suffocated in his sleep.

Tim's eyes wandered back to the figure of his mother who sat on the edge of his bed speaking of the wondrous adventures of a man named Hannibal. But, even as she spoke he found himself longing for his father. He missed watching his pale blue eyes through the thin layer of dust that always coated his spectacles. He missed the warmth and wrinkles of his smile and the feel of his whiskers when they hugged.

His mother stopped talking."You're not listening," she said softly. "Tired?" Tim nodded and closed his eyes. He wasn't really all that tired, but he needed to be alone right now. He felt his mother's lips brush his cheek then heard her whisper in his ear. "I miss him too, honey."

She hugged him for a long moment, clinging to her dead husband's legacy and aching for his love. Finally, she rose and went to the door. "It's okay to remember, Timmy. It's always okay to remember."

But, that was then. Now, with the irony of those words still ringing in his ears, Tim was eighty-two years old and sat quietly on a bench, wondering where the hell he lived and watching a bus pull to a stop in front of him. Air escaped as the doors opened. The driver leaned forward on the steering wheel and smiled at Tim.

"Hope you haven't been waitin' too long."

"Nope," Tim replied. "I don't think I have. What time is it?"

The driver looked at his watch. "About six-thirty. Sorry. Guess I'm runnin' a little late."

"That's okay," Tim forgave him as he got to his feet. "I've gotta be gettin' home." Tim grabbed the bag holding the week's earnings and looked around, turning in a circle as he did so.

"Well, let's go, friend. Time's a wastin' ." Tim agreed and, dazed, climbed aboard.

"Don't worry about the fare," said the driver, putting hand over the the coin box. "I'm the one who's late. It's on me."

Tim O'Sullivan thanked the man and sank into a side-seat near the driver. He sighed heavily as the Airport Express roared out of Hillyard and headed for Geiger Field.

Tim watched the highway. The broken yellow divider-line whipped past at a steady clip. Farmland spread out on both sides of the bus as the setting sun bathed the road in fading warmth.

He watched cows grazing, heard a car horn honk and saw young horses cantering within a fenced meadow. Red barns dotted the landscape between wide-open fields of alfalfa, wheat and barley.

The horn honked again and a car passed, the driver waving to the bus driver. Flocks of crows scurried from one field to another—ignoring the occasional scarecrow. Every now and again Tim saw a rabbit poke his nose above a roadside ditch then duck or scurry off as the bus roared by.

His stomach growled, it must be nearly dinnertime. His wife would be getting hungry. The bus pulled to the side of the road and slowed to a stop.

The delay annoyed him, but he tried to be pleasant as the bus doors sighed open and a middle-aged woman climbed aboard.

"Good evening," she said to the driver then turned her attention to Tim. "Hello, Timothy."

"Hello," Tim responded, trying to figure out who she was. He was pretty sure she was a customer of his hardware store, but her name escaped him for the moment. There were a lot of things that escaped him for the moment. This was disconcerting and annoyed him all the more.

"Where're you going, Tim?" she asked.

"Home," Tim replied.

"I thought so," she said, holding out her hand. "Come on, I'll take you there."

"She's taking me home," Tim said, nodding to the bus driver.

"That's right partner," said the bus driver exchanging looks with the woman. "I'll be going the long way, and this nice lady can get you there a whole lot quicker than I can."

Tim could hardly believe what he was hearing. "You're gonna take me home right now?" he asked.

"I'll have you there in ten minutes," the woman promised.

Relief that he was finally going home was overwhelming and he almost broke down and cried. Fighting back tears of gratitude, he rose from his seat and followed the woman to her car.

She made a wide U-turn in the middle of the highway and started back to Hillyard. "I was just coming out of Rosauer's after doing a little grocery shopping and I thought it was you I spied climbing aboard that bus." She smiled at Tim as he turned his attention to the landscape whizzing past his window.

When the car finally pulled up in front of the hardware store, Tim's heart fell. "I just left here," he said, annoyed.

"This is where you live, Tim."

"Oh, for Christ's sake," Tim said, throwing the car door open. "I should know where I live. This is where I work."

Tim stepped onto the sidewalk still holding the paper sack with the week's earnings in it.

"Timothy." Tim looked around. The voice seemed to be coming from overhead. "Timothy!" He looked up. His wife, Bonnie, was leaning out the window of their second-floor apartment. She waved and Tim waved back hesitantly.

Sticking his head back into the car,, he said sheepishly, "I guess I do live here. Thank you Miss—"

"Irma." The woman smiled warmly.

A memory rushed back to him. "Say, I recognize you. Aren't you one of the Worthy girls?"

"Used to be. It's Bridgeport, now."

"I think I used to know your grandfather when I was a little boy."

"Could be."

Soon after that, Timothy O'Sullivan realized it was time to sell his hardware store. Bob bought it, and changed its name to Bob's Hardware Store.

I busied myself surveying the store's inventory while Irma watched Bob fill her order. Bob, a yellow pencil still parked behind his ear and blue bib apron secured around his neck, worked quickly, wearing a smile and sweating profusely as he packaged up nails, hinges and such. He tied each paper bag closed with a length of white string and labeled the contents with a grease pencil.

When Irma and I returned to the step van, the cartons of newly bought eggs that had been sitting on the floor—all the while being warmed by the sun—were askew. At first, I thought it was my imagination, but I thought I heard a high-pitched chirp. Upon closer inspection we discovered baby chicks stumbling out of the egg cartons and a few already freed from their shells, scampering over the floorboards.

"Oh my word," Irma gasped.

I was delighted. "What are you gonna do with them?" I asked.

"Well, they're a little too far along to make omelets out of," Irma quipped. "I guess God must have meant for me to have them, but I'll tell you, I never thought of myself as a chicken farmer before now."

"Maybe we could name one of them, Peabrain," I suggested as we were heading back to the farm.

"That's as good a name as any," Irma agreed.

When we pulled off the main road and followed the winding drive up to the farmhouse, Uncle Harold waved from a tractor that had seen better days. He'd traded it for some carpentry work and after a little fixing-up the old tractor had never run better.

Rich, black soil was being turned over and Duke was sitting on the seat next to Harold, grinning. Harold was whistling a tune that he made up as he maneuvered the oversized steering wheel with uncommon alacrity. I remembered Uncle Harold once telling me: "If life were a jelly doughnut, I'd probably eat it." Duke jumped down from the moving tractor and ran barking to the step van.

Later that same day, Harold drew up plans for a new chicken house. The construction would take almost two years to complete—elaborate wooden coups and perches for the mother hens, ultra-violet lights to keep down the spread of bacteria, linoleum flooring, baseboards of sheet metal, and wood paneling sealed under several coats of white enamel for easy cleaning—every amenity would be employed to keep the chickens comfortable, clean and healthy. But, for the time being the old coup with its collapsing roof would have to do.

The following summer, Nellie and I obliged Irma's bidding and waded through the moving sea of yellow chicks to fetch more eggs before we went fishing. We shuffled our feet for fear of crushing the babies and spent more time gathering them up in our hands and petting their soft down than we did collecting eggs.

The hens sat in their nests clucking with disapproval and attempted to peck a hole in my arm whenever I slid my hand under their warm bellies to feel for an egg. But, I was swift and managed to move out of their reach most of the time.

When we had finally gathered enough eggs and carefully placed them in the basket Irma provided, we walked to the water pump and rinsed our hands. The water was lukewarm and clear.

Later, after dropping off the eggs at the house, we found Paul behind the barn with his pants down to his knees peeing on the alfalfa.

"How embarrassing," I said. Nellie giggled.

Paul whipped around, jerking his pants up and zipping his fly. "You shut up," he sputtered. He flushed when he saw Nellie standing there, staring. "What do you two want anyway?"

"We're going fishing," I announced.

"Yeah, I know. Irma told me when I went down to the house. I guess I'm supposed to go with you. Well," said Paul, "I ain't gonna stand around here all day yakin'. I've got better things to do. Let's get this thing over with."

When we reached the woods, Nellie snapped a long branch off a willow tree and handed it to me. "This is your fishing pole." She tied a piece of string to one end and showed me how to balance the pole on my shoulder.

When we left the main road, we stepped into the woods of ash and pine and traveled along a beaten path of trampled weeds and undergrowth. The day was warming and the air grew pungent with the scent of fireweed and honeysuckle. Paul followed me while Nellie led the way.

Occasionally, I would spy a miniature wild iris or daisy growing amongst the weeds and stop to pick it.

"They're just gonna wilt by the time you get back," Paul informed me.

"We'll put 'em in water when we get to the pond," said Nellie. I found an old tin can along the way and stuck the flowers in it.

"This is the best spot to fish," came Nellie's hushed voice when we settled down on the bank next to the pond.

"You have to be very quiet when you fish."

I silently tied the loose end of the string around the worm Nellie offered and dropped it into the water.

"You're gonna need a hook if you're gonna catch any fish," sneered Paul, leaning against a tree.

"Shhh," Nellie warned, putting her finger to her lips. "You have to be quiet when you fish."

"Why?" I asked in a whisper.

"That what my Dad told me and he fishes here all the time, so he oughta know."

We sat in the shade under a tall tree with large branches that dipped low over the water. Bushes lush with purple berries surrounded the pond. The sun rose above the trees and shone bright while a wind whispered through the leaves and made ripples on the water. Black insects with elongated bodies and stick legs skated on the pond's surface. They moved with quick jerks and made sudden stops. "They're called 'skippers'," Nellie said.

We sat quietly. Nellie and I, basked in the filtered sun and watched nature unfold in the early morning hours. Paul stomped back and forth impatiently, chewing on blades of grass and violently tossing stones into the dense forest where pockets of disturbed wildlife dived for cover. Nellie had brought a sandwich along and offered me half.

"You want some?" she asked Paul.

"Nah, I'm not hungry. What kind is it?"

"Peanut butter and Miracle Whip."

"Forget it," Paul scoffed.

"It's good," I said and ate my half greedily. "Do you think we're gonna catch any fish?" I asked Nellie.

"Not without a hook you're not," warned Paul.

"Fishing's got nothing to do with catching fish," Nellie said simply.

"What's that supposed to mean?" Paul asked, chucking another rock into the woods.

"That's not what fishing's all about," said Nellie leaning back against a tree. She puts her pole down and closed her eyes while her line dangled in the water.

"What is fishing about, then?" I asked.

"Fishing is sitting by the pond and feeling the sun on your face and listening to the birds and the bugs and hearing the water move. That's what fishing is. It's got nothing to do with catching fish." I nodded as if I understood. "Besides, there ain't any fish in this old pond anyway. There probably never was as far as I know."

"I don't believe this," Paul sighed with disgust. He kicked at the forest floor and pine needles shot into the air.

On our way back to the farm, I stopped and picked more flowers. By the time we reached the farmhouse the tin can was stuffed with weeds and wild flowers.

Uncle Harold was in the living room, as were most of the grownups, reminiscing about the past, complaining about the present and casting doubt on the future. When he leaned back in his chair the wood groaned. He folded his fingers behind his head and gazed out the window as the crowns of Paul and Nellie and I bounced into view.

Brother was talking, rambling on, telling another one of his war stories, this time about his near-death experience in France. His voice droned on, losing Harold's attention—he'd heard this one so many times he could have told it himself. He watched the cousins approach the hill's summit. I was carrying a bouquet in one hand and a fishing pole in the other. Nellie kept one hand protectively on my shoulder. Paul brought up the rear.

"What's that Theresa up to?" Paul asked.

Ahead of us, Theresa squatted on the lawn, her blue dress spreading about her like a funnel. She was concentrating on an object in her hand. The dew had evaporated and left the grass dry and warm to the touch.

"Come on. Let's go see," said Paul, rushing ahead.

Nellie and I followed at a distance. Paul hurried to Theresa's side and was soon squatting next to her, his gaze fixed upon the object in her hand.

As I came closer, I thought I heard a faint humming or buzzing. Philip and Steve, who were busy building a fort under the belly of an old combine, dropped what they were doing when they saw a crowd gathering and joined their cousins. "Well," said Steve, his hands stuffed deep into his pockets and rocking back on his heels, "what have we got here?"

"Yellow jackets," Theresa announced without looking. She was pressing hard, holding a jar over a hold in the grass. A hornet's nest. A seemingly endless army of stinging wasps streamed out of the earth only to find themselves trapped in a glass prison.

"There must be a million of 'em," said Philip squatting too.

"At least," Paul agreed.

"They're coming out of that hole in the ground," I whispered to Nellie. Nellie just nodded and watched the jar fill.

"Jesus, they just keep comin', don't they?" said Philip, grinning.

"They're packed in there like sardines," Paul observed.

There was no more room for the hornets now. They were crawling over one another, their shiny bodies pressing against the walls of the jar while still more attempted to force their way in. The thousands of black and yellow-striped wasps melded into a single entity—a solid mass of squirming, dangerous insects slick with condensation. The air inside the large mayonnaise jar was hot and alarmingly thin in oxygen.

"They look like their sweating," I whispered to Nellie.

From their undulating abdomens, needle-sharp stingers probed the cramped air, like soldiers brandishing their swords before battle. The ones near the top of the jar had grown angry and began to sting the wasps in close proximity. The buzzing and the humming grew in intensity.

"Unless she plans on staying in that position for the rest of her life, sooner or later she's gonna have to let go of that jar." Nellie took my hand and pulled me away. "Anyone who hangs around here has gotta be crazy."

I let myself be led away. "Come on. This ain't no time to dilly-dally," Nellie said, keeping her voice low.

"Where're we going?"

"As far away from here as we can."

Time and again I looked back. Theresa was still squatting on the ground, her dress smudged with grass stains, her hand holding down the jar. If she were to let go, I thought, the jar would go sailing off into the sky, jettisoned by the sudden burst of buzzing wings powered by a million angry yellow jackets.

I imagined it would take every ounce of Theresa's strength to hold the jar snug against the earth, her arms trembling from the effort. Beads of sweat would drip from her hairline and slip under her eyelids and sting her eyes, blinding her vision.

"I don't want you thinkin' I'm just being a bossy-pants, Benny," Nellie informed me as we fled across the lawn. "But Irma's gonna be awfully mad at me if anything happens to you. And, I'd rather die than disappoint her."

Paul took little notice of me and Nellie's withdrawal. The yellow jackets were far more interesting. Their buzzing was changing in pitch and their obvious misery was a delight to behold.

"Let me hold the jar for a minutes," he said.

"No thank you," was Theresa's reply.

"Aw, come on. I just want to hold it for a minute. Then, I'll give it right back to ya."

"Forget it, Paul. It was my idea to begin with." Theresa could feel the vibrations of the yellow jacket's buzzing through the glass and it tickled her hand.

"You know, I heard their stingers are so sharp they can actually poke right through the glass," Steve teased.

"Oh, they cannot," Theresa scoffed.

"Yes they can," said Paul, playing along. "If their bodies get real slick and they start to sweat, they can work their stingers through the invisible air holes in glass."

"Slicker than snot, they can," Steve added.

"You boys are so full of it," said Theresa. Boys were always ganging up on the girls and trying to scare them. But, the funny vibration that was working its way through the glass was no longer tickling, but irritating, now. If there was even a grain of truth to what they said, the idea of hanging onto the jar much longer was about the last thing she planned to do.

Uncle Harold accepted a cup of coffee from Irma, took a sip and smacked his lips appreciatively. As he placed the saucer on the windowsill he saw Nellie and I hurrying away from the gang of cousins.

At first our scurrying seemed almost casual, but when a shout went up and he heard Theresa scream, Nellie and I broke into a dead run.

"What in the dickens is going on our there?" Brother asked, risking to his feet.

"I'm not sure" said Harold, "but Benny and Nellie just skedaddled to the rear of the house. And, rest look like they're high-tailin' it any direction they can."

"Hell bent for leather, I'd say," said Brother.

"Kinda looks that way," Harold agreed.

When the parents realized whaat had happened and upon seeing the welts of bee stings covering their children's bodies pandemonium erupted, It was through Irma's coolness and wise intervention that they weren't all hauled off the nearest hospital.

"I know the best thing for bee stings," said Irma heading out of the room.

When Irma entered the kitchen she headed straight for the spice cabinet. Nellie and I were seated at the kitchen table.

"Well," she said, pushing bottle aside and finding what she was looking for, "looks like we've got at least two young people here with good sense."

"I'm no fool," said Nellie and smiled at me.

"I know you're not dear.," she said, tightening her grip on a narrow necked bottle and heading for the door. She stopped suddenly when she reached the table. My what beautiful flowers!" She exclaimed, pausing to examine the bouquet.

"Benny picked 'em for ya," said Nellie.

"Nellie helped."

"Well," she said, "thank you, both," giving us each a hug. "How very sweet."

"Watch got there?" Nellie asked, nodding to the bottle Irma had grasped.

"Vanilla extract," said Aunt Irma, straightening. "It's the best thing I know of for bee stings. Why, the smell alone can stop a tear."

For years after the bee attack, Paul blamed himself for the catastrophe. If he hadn't gone along with Steve and teased Theresa, she might never have panicked and dropped the jar. I knew it wasn't Paul's fault, but never told him. Paul was easier to get along with once he had to live with the guilt. It seemed to humble him just enough to make him more human.

Since I had my back turned and was running away with Nellie, I never saw what really happened. I learned the truth one day when I overheard Lois, Theresa and Steve's mother, telling the story to my mother, Joanie. They were seated at the kitchen table, where my mother entertained nearly all her guests.

"Well, pretty soon," Lois was saying, "Theresa's jar fills up—I mean, it was packed to the brim. She couldn't have stuffed another yellow jacket in there if she'd wanted to. And, they're all buzzing and squirming and crawling all over each other, trying to get out—it was probably blocking off their air supply too, for all I know. Well, they started buzzing so hard she could feel the vibration through the glass and she said it tickled.

"Now, along comes this other yellow jacket. He'd been out hunting when Theresa got this bright idea. Now he can't get back into his nest, 'cause Theresa's got a jar over it. But, he tries anyway and Theresa keeps shooing him away.

"Well, before you know it, he gets mad and flies right up her dress and stings her on the butt. Theresa jumped up hollering her fool head off and naturally she knocked the jar over when she did.

"About that time a thousand yellow jackets decided to teach them all a lesson. When I walked out on the back porch all I saw was Theresa running away just as fast as her little legs could carry her—screaming to Livin' B. Jesus—and waving her arms about like this."

That was last summer. Now, the hollyhocks were in full bloom and the bumblebees that flew in and out of the enormous blossoms were so fat they had to take running starts to take off. Again, an overweight bee failed to reach enough speed, tumbled helplessly over the petals and fell to the earth.

I knelt and placed another twig in its path. It climbed aboard willing enough, but then froze as I lifted the twig above my head.

"Come on, now. You can do it. Fly," I said, jiggling the twig. The bumblebee buzzed its wings, lost its balance and fell to the ground again. Yep—the bees weren't nearly as aggressive as they were last summer.

Maybe it wasn't the bees, I thought. Maybe it was me. Now that I was five, I was aware of things that were a mystery just a year ago. Secrets. Family secrets that made my mother weep.

Sometimes I'd find her studying an old photo album. I'd watch her eyes skim over the images of a little blonde-haired boy playing in the snow or sitting on a porch step.

Suddenly, she'd begin to cry. Without making a sound, tears fell onto the photographs then rolled off and the black pages soaked them up like a giant blotter. I would even attempt to console her by putting my arms around her, but it would embarrass me that she shook so when she wept.

The morning sun warmed the back of my head and when I touched my hair it felt hot. I was squatting in a gravel alley trying to coax another bee into flying. The bumblebee was lying on its back waving its feet in the air, running a marathon to nowhere. A strange and sad sight, I thought.

Tomorrow would be the family picnic. I was looking forward to the long car ride and spending time at Harold and Irma's. I missed my cousins and the woods.

But, something else was going on. I wasn't sure what, but I could tell by the look in my mother's eyes that something was amiss. I suspected it might have something to do with another one of those family secrets. There were a lot of them. Like that night just before last Christmas.

It seemed like only yesterday when it was winter, and snow still covered the earth. It appeared as though nature were attempting to warm the planet by throwing a big white blanket over it.

At dusk, when I walked out on the back porch, the winds that drifted down from Canada made me shiver and rub my arms for warmth. If I looked up I might see convoys of mallards or geese making haste in their journey south.

The robins, the jays and the sparrows had long since disappeared with the last chilly days of October. They left behind their round-bellied cousins, the chickadees, who spent their days flitting about, hopping from one naked branch to another searching for snowberries.

They were especially fond of the trees on the corner of Thirteenth and Maple, and attended their banquet with whistles and chirps that could be heard for blocks. There, the fruit-laden branches were so heavy they bowed painfully to the ground like an aged tap dancer accepting applause. When the birds added their weight, the branches nearly snapped from the burden.

I, however, took little notice of the avian hostilities, though their strident chattering was enough to keep my baby sister awake during nap time. Instead, I stood in the foyer with my face pressed against the glass in the door watching the snow fall. It seemed like only yesterday when it was autumn and I stood in the front yard with a maple-seed pod pasted to the bridge of my nose, watching the leaves fall from the trees. I'd hurl myself onto a mountain of leaves and lay on my back to watch the dizzying flight of seed pods spiraling to earth. Sometimes I imagined myself piloting one of those whirling pods, much like a cowboy might cling to the back of a bucking bronco.

But, those days were gone. Now, it was winter.

Across the street, the vacant lot that was alive with grasshoppers and flying insects in the summer was now a white mantle of virgin snow, untouched and desolate.

In the distance, though barely discernible, I could hear the children of the neighborhood congregating on the crest of Comstock Hill, preparing for their daring sleigh rides.

I sneezed then wiped my nose with my sleeve and sighed. If only I wasn't sick. There were times when I envied Alan Borgy, my best friend, his freedom, and in those moments I hated his guts.

The snow fell steadily with almost no wind; giant flakes floated past the streetlights like popcorn. I could hear a car approaching when it was still several blocks away, the tire chains making their own special music, growing louder as they neared, clinking and humming, pounding the packed snow as they rolled past.

From listening to "The Adventures of Superman" on the radio, and now with the Superman series starting its run on television, I convinced myself that I, too, was Superman. When I wasn't playing Cowboys and Indians I was running down the sidewalk with my arms raised over my head making the sound of Superman flying— which sounded a lot like a strong wind seeping through a narrow opening.

I poked my finger under my shirt and felt the raised "S" on the Superman costume I wore under my clothes. If only my parents had had the foresight to name me Clark when I was born, I thought, as I adjusted my glasses.

I had perfect eyesight, but insisted upon wearing glasses anyway. I'd punched the lenses out of a pair of sunglasses and just wore the frames. They were the perfect disguise.

Alan Borgy had actually dropped to his knees the day I confessed my true identity and pulled up my shirt to reveal my big red "S". Wearing a full costume under my regular clothes could be very uncomfortable at times, especially on hot summer days, but I got by and even managed to coax my red cape down one pant leg, though it kept bunching up behind my knee.

I reasoned I must have ventured too close to some Kryptonite at one time—which would explain why I couldn't really fly. Someday the effects would wear off— I hoped.

Alan Borgy broke down and wept upon hearing of this, having satisfied himself that it was the Kryptonite in his watch—the stuff that made the numbers glow in the dark—that was to blame. He promptly crushed the watch under a rock and buried the remains. He told his parents he lost it.

Alan Borgy's parents had moved to the U.S. from Germany sometime prior to the war. I suspected the entire family of being Nazi spies. The evidence was overwhelming—Alan Borgy's father sported a Hitler mustache, spoke with a German accent, and to top it all off, his first name was Adolf.

I promptly alerted my father to the danger. But, my dad only laughed and scooted me out of his study, confusing a mind that was filled with the images of Auschwitz, Dachau and Buchenwald— pictures that filled the pages of the dozens of books my parents placed in my lap.

"Kryptonite," I whispered. The thought had suddenly occurred to me. Kryptonite can harm Superman. That's the only thing that could have given me a cold and destroyed my plans for playing in the snow. I squinted through the thick glass in the front door and made it fog when I coughed.

My mother paused from preparing dinner; fresh-cut onions and carrots sizzled in a thin pool of butter. She sprinkled it lightly with salt before putting the lid on, then turned down the heat. The savory fragrance snaked through the house.

In the living room, the television cast a bluish light upon the walls and on the screen a little boy's face came to life on the label of a jar of peanut butter. He broke into a smile. "Sunny Jim?" he said. "Gee, that me!"

We were the first on our block to own a television. It was so heavy, it took two burly men in blue overalls twenty minutes to maneuver it out of the truck and into the house.

The men huffed and puffed as they struggled with the hand truck. The caster wheels left deep impressions in the carpet as it was rolled across the room. With smiles and a flourish, they slit the cardboard box open and lifted the impressive new addition onto the spot my mother had reserved.

A large, cumbersome Packard-Bell with a cabinetry of blonde wood and hinged doors, swung open to reveal a massive 12-inch round green screen. After the men left, I stood in the living room with my hands on my hips in awe. A real television— just like in the science-fiction movies, with knobs and dials and a huge screen. The possibilities were ominous.

"Can you talk to it?" I asked my father.

"You can talk to it if you want to," he answered, "but I doubt if it will hear you." My dad plugged the set in and fell back on the couch; I snuggled up next to him. "Let's see what's on."

An image seeped slowly into view until it was bright and unwavering. It was the profile of an American Indian with numbers circling his head and lines running diagonally to all four corners of the screen.

We sat staring at the test pattern for some minutes before my dad finally asked, "What do you suppose that means?"

I wasn't really sure, but I had a pretty good idea. "It's the introduction. It means they're gonna be showing a Cowboy and Indian movie pretty soon."

"You think so, huh?"

"Yep," I nodded with conviction, folding my arms. I settled back comfortably on the davenport and waited for the movie to start. My father waited patiently with me. Several minutes passed and the same image stared back at us.

"How long do suppose this is going to take?"

"I don't know," I admitted. "But, it shouldn't be very long. I hope it's a Hopalong Cassidy movie. I just love Hopalong Cassidy."

The Cowboy and Indian movie never did appear as I had predicted. In fact, we had bought our TV a month before the local television station began broadcasting its first real programs. For some time after that I felt a wave of embarrassment whenever a test pattern appeared on the screen.

But, the memory of personal humiliation soon faded and in time the television became the family gathering ground. Especially at six o'clock on Friday nights when Pabts Blue Ribbon and Gillette Blue Blades brought the Fight of the Week into the nation's living rooms.

In 1952 Rocky Marciano was the Heavyweight Champion of the World and he found no greater admirers than in our living room. I sat on the floor as close to the television screen as was humanly possible. When my mother came into the room she'd make me scoot back, warning me that I might go blind if I didn't.

I sat cross-legged, like an Indian and ate popcorn and drank strawberry Kool-Aid and watched in wonderment as two perfect physical specimens attempted to knock the crap out of each other.

My father slipped off his shoes, stretched his legs out in front of him, crossed his feet at the ankles and sipped beer. "Why don't ya hit the son of a bitch? You jerk!" he shouted at the television.

My mother sat on the edge of the couch and leaned forward with both fists doubled-up. She helped the champion by swinging at an invisible opponent and shouting encouragement. "Hit the bastard! Knock the prick flat on his ass!"

This was usually a grand time for me. I took great delight in these family gatherings and was especially pleased when my mother and father joined me in front of the television to share in the festivities.

But, this wasn't Friday. This evening I stood in the foyer with my nose pressed against the glass in the door watching the snow fall and listening to the children far in the distance getting ready for their sleigh rides.

"Well, as much as I'd like to, I guess I can't keep you cooped up indoors for rest of your life, can I?" I agreed with a nod. "Well, come on, then," said my mother taking my hand and leading me away from the door.

Upstairs in my bedroom my mom pulled cold weather gear from the closet and helped me dress.

"I know you've got a cold, but it seems to me if you cover up real good you'll be okay. A little exercise won't hurt you as long as you don't over do it. Besides, unhappy children never get well as fast as happy children do. Did you know that?" I shook my head. "Well, it's true. My father told me that—your grandfather."

"Grandpa Worthy?"

"Yes, that's right" She knelt on the floor while I sat on the edge of the bed. She smiled up at me and brushed my hair away from my eyes. "I don't think staying in all day feeling sorry for ourselves is very good medicine at all. Do you?" She pulled on my boots and buckled them. "It's not like we're living in Antarctica, is it?" I shook my head and thought she was doing a pretty good job of convincing herself that it was safe for me to venture outdoors.

"Come on." As she led me downstairs, she wrapped a muffler around my neck and pulled my cap down over my ears. "If you start feeling tired I want you to come right back in." She stuffed a hanky in my pocket. "Colds can get nasty sometimes."

I turned and looked through the glass at the snow. She held me for a moment, embracing me while my back was turned. She wanted to tell me she loved me, but something stopped her, a feeling that it would be unfair to say those words to me while not being able to speak the same words to Jeffery, so she kissed my cheek instead.

Outside the snow was still falling. While my mother had helped me dress, I feared the world, or God, or whoever was running things would know I was coming and deliberately stop the snow from falling just for spite. But, He hadn't, and the flakes had grown even larger and clung to my eyelashes when I looked up.

I tried building a snowman. It was hard work, but the consistency of the snow was perfect. When I trudged through the mantle, it gave under my boots in dull whispers.

I scooped up a handful of snow and packed it into a firm ball, the ice crystals wuffling as I forced them into a tighter space. I dropped it to the ground and began rolling. Before long the snowball grew to the size a boulder—so large it couldn't be budged. It was nearly as tall as I was and impossible to lift.

A green Buick belching exhaust pulled to the curb; the snow groaned under the tires as it came to a stop. My father stepped out, pulling his briefcase after him and waved at me as he ran for the cover of the house. Moments later he reappeared and stood in the doorway, a small sled under his arm and our dog Cindy, a black Scott Terrier, at his heels.

I took my father's hand and we headed for Comstock Hill, pulling the sled behind us. Cindy ran ahead. Occasionally she stopped to attack the ground, first snapping then finally taking a bite out of the snow.

Dinner wouldn't be served for another hour or so and already the sky was turning dark. We trudged up the hill until we came to Cedar Street.

Cedar was one of the town's busiest roadways, but after a heavy snowfall, it became treacherous for driving. After a number of serious mishaps, the Mayor ordered the street closed and, to the delight of the children, it became the neighborhood sled run. White sawhorses straddled black smudge pots at each intersection to warn motorists away.

The smudge pots looked like cannon balls with their fuses lit. Their flames danced to the whims of the wind as the oily smoke twisted away, turning parts of the sawhorses black.

As we neared the crest of the hill, shouts of excitement and laughter filled the air. It was a carnival of life. Boys and girls of all ages, bundled in coats and mufflers, wool hats and earmuffs ran about wildly, some falling down deliberately and rolling in the snow.

I joined a circle of children from my neighborhood. Alan Borgy was there, so was Dorothy Studebaker. Mark Morey came trotting up at about the same time.

"Hey, Mark."

"Hey, yourself," Mark responded.

I cast my eye to the thin pale line that marked the horizon. In the winter, sunsets were benign, it just got gradually darker, like a light bulb being dimmed. Street lights began turning on and we watched the snow that still continued to fall float past the lamps.

We stood quietly together looking at the town below. Sometimes I had a tough time telling the difference between what was real and what wasn't. The scene below looked remarkably like a lot of paintings I'd seen; I wouldn't have been at all surprised to see a sleigh pulled by eight tiny reindeer suddenly lift from one of the roofs and take to the sky. Mark was the first to break the silence. "Hey, you guys want to share a cigarette?"

I looked at Mark incredulously. "You've got a cigarette?"

Mark pulled a non-filtered Camel from his pocket. "My brother hides them under his pillow. He won't even know it's missing." Mark looked around conspiratorially. "Even if he did—who's he gonna tell?" I nodded, agreeing with the logic.

Dorothy, whose mind was elsewhere, suddenly noticed she was standing alone and quickly joined the three of us as we gathered at the edge of the hill.

"How come you didn't invite me?" she asked, trying to sound hurt.

"We thought you were trying to quit," Alan quipped, smiling.

"You sure the parents won't notice?" Dorothy asked.

"Nobody will notice," Mark assured her as he fumbled with the matches. He took a puff and passed it on to Benny. "Anyway, they won't be able to tell if it's smoke or our breath in this cold. For all they know we could just be breathing hard, like we're out of breath or something."

We stood with our backs to the crowd sharing the cigarette and looking out over the sprawling town. To our left, the gully loomed.

"Hey, Benny," said Mark, nudging me. "Remember last summer when we . bought a whole pack of Camels?"

"A whole pack?" Alan asked with disbelief. I remembered. How could I forget?

It took Mark and I most of the morning visiting vacant lots, wandering up and down alleys, and checking along the roadside before we found enough pop bottles to add up to thirty-five cents. Finally, when we figured we'd collected enough, we started for the nearest grocery store.

We took turns pulling my red wagon, the bottles banging against one another singing a sweet song on that warm summer morning as we pulled the wagon down the sidewalk.

Mark's parents owned a tiny grocery store that sat in the middle of the block. It was attached to their house, and sometimes when I was buying Popcicles I could hear doors opening and closing inside the living quarters and once in a while I'd even catch a glimpse inside. But, in all the years Mark and I were friends, I never ventured into their domain, partly because I was never invited, partly because Mark had two older brothers.

In front of Mark's house/grrocery a monstrous chestnut tree loomed in the of an untrimmed hilly lawn. The lawn was divided by a concrete walkway that led to a small, white storefront with large windows and a screen door.

The walkway was cracked and shattered where the tree's overgrown roots had snaked beneath and lifted the sidewalk into a precarious chevron. The bottles sang merrily as we maneuvered the wagon over the little mountain.

For a reason Mark couldn't seem to remember, his parents were gone, leaving his two older brothers: Larry and Jerry, to mind the store.

Jerry, eighteen, was sleeping in that morning—he'd been out late the night before cruising the drag with his girlfriend until two in the morning. So Larry, sixteen, was standing behind the counter drinking an orange pop and reading an Archie comic book when Mark and I entered the shop.

Larry read Archie comics because he loved Betty and Veronica. They had even made brief appearances in his dreams. Betty was cute, with nice round breasts, but she was too innocent. Veronica, on the other hand, was beautiful, had big pointy tits and definitely got around. It was less than a week ago Larry had the pleasure of a nocturnal emission when, in a dream, he'd pulled up Veronica's blouse and to his elation she wasn't wearing a bra.

The little bell over the door rang nervously as I pulled and Mark pushed the little wagon into the store.

"We need some cigarettes," I said, my heart pounding.

Larry put down his comic and took another swig of his orange pop.

"I mean my mother needs some cigarettes." I searched my pockets for the note Mark and I forged earlier. Mark stood behind me and sheepishly examined a shelf of canned goods, trying to blend into the background. Finally, I found the note and handed it across. Larry eyed us both suspiciously as he unfolded the note.

Gentelmans:

Please to let my liddle boy Benny by me sum cigerets. Heres sum bottles for them. Thanks to you a lot.

Your Frenly Cusomer,
Mom

Larry read the note with some interest, took another swallow off pop and sized up his little brother and me. I could feel the palms of my hands grow wet. The only sound in the store was the whirring of the fans and motors coming from the refrigerators.

Suddenly, startling us both, Larry asked, "what kind does she smoke?"

"Who?" I asked, unprepared for the question.

"Your mother," said Larry, waving the note.

"Oh, yeah," I said putting a finger to my lips. My eyes skimmed over the dozens of different multi-colored packs neatly stacked and displayed next to the cash register. They all looked so good.

"Camels," Mark whispered, shading his mouth with his cupped hand and breathing the word into my ear.

"Camels," I said.

Larry suppressed a grin and pulled a short, firm pack from the stack and slapped it onto the counter. Our eyes feasted upon the delicacy. Mark fairly drooled, saliva spilling down his chin and wetting his shirtfront.

"Thirty-five cents," said Larry.

Mark and I pulled the bottles out of the wagon and lined them up on the counter. Larry caught my eye.

"My daddy didn't leave my mother any change when he went to work this morning," I lied.

"Okay," said Larry looking at the dozens of bottles, most of them covered with dirt and cobwebs. "Does she need matches?"

"Matches?" Mark slapped his forehead. It was incredulous they hadn't considered how they would light the cigarettes—assuming they'd actually succeed in acquiring a pack.

"Yeah, matches!" I said realizing how badly they would be needed. While Larry looked for matches, Benny watched an ant try to crawl up the inside of a bottle. Invariably, it slipped on the smooth glass and slid back to the bottom, only to keep trying over and over again.

Outside, the Camels successfully procured, Mark and I, renegades of the non-smoking world, self-appointed disciples of the nicotine God, raced down the sidewalk, pulling the wagon behind us until it flipped. Unstoppable, we dragged the overturned wagon on its top, sparks shooting. We finally discarded the wagon in my garage before running down the hill to the gully.

The gully was my region of enchantment. It was as if God had taken His thumb when the Earth was still soft and warm and pressed it into the land, leaving a deep impression—a dimple in the middle of the city where evergreen trees, crab grass, buttercups and purple iris grew wild.

A section of the gully had been duly christened: "The Marble Mine." In reality it was an abandoned Indian graveyard disturbed by the railroad when the tracks were laid. Periodically, the earth belched up chunks of marble of various colors and sizes—the remnants of broken tombstones.

The tracks ran adjacent to the gully. At night, when the trains rolled into Spokane, they blew their whistles and the sound bounced off the deep slopes, echoing back and forth like the inside of a bell.

At the base of the gully, Mark and I straddled two large fallen trees and divided the cigarettes between us.

Puffing on our tobacco sticks, but never inhaling the smoke, Mark and I consumed the entire pack during the next two hours, discarding our butts, some still smoldering, onto the forest floor.

Between puffs, I told Mark of my love for writing and tried to express the feeling I got when I put words on paper. I must have failed miserably because Mark answered me by telling me of his desire to become a steam-shovel operator someday and grow a beard. When we were finished, we trudged back up the steep slopes of the gully, leaving behind the empty pack of Camels and the spent matches. We retired to my garage.

I immediately set about greasing my vehicles. Mark looked on while I explained why you should only use creamy and never chunky peanut butter on your wheels. And why you should only do it in the summer when it was warm because in the winter the peanut butter would harden like cement and lockup your wheels. I also explained how important it was to clip playing cards onto the hub of your trike. New playing cards were best and wood clothes pins worked fine for securing them. The sound of the slapping cards made the trike sound "neat," and the peanut butter made it easier to pedal.

My dissertation was abruptly interrupted by the sudden appearance of a fire truck tearing around the corner just yards from where I knelt. Its sirens screamed and the engine roared as the driver punched the accelerator and the hook 'n ladder sped down the hill towards the gully.

Mark and I dropped what we were doing and joined our neighbors at the gully's crest.

"What's all the excitement about? I heard someone ask. Someone else answered, "fire," and a few neighbors groaned. My eyes fell to the two logs on the floor of the gully.

"Look," someone said, "somebody must have started a fire down there and it spread up the side of the gully." The ground was charred black. It was plain to anyone the fire had indeed started near the two fallen trees. The flames had burned their way up the once grassy slope leaving a black trail in their wake. It consumed a garage and blackened half of an attached spilt-level house before the fire department managed to bring the fire under control.

I suddenly felt ill.

"Are you okay?" It was Dorothy Studebaker.

"Yeah, I'm fine."

"You don't look fine," she said. "If you're Superboy, why don't you fly down there and blow the fire out?"

I studied Dorothy through the pop-bottle lenses of her horn-rimmed glasses. "Because, stupid, if I did that, everyone would know my secret identity."

Yeah, I remembered last summer alright. The guilt of knowing I caused the fire gnawed at my insides like a wild animal eating its prey. Sometimes the memory of the damage fell upon me without warning and made me feel like I was coming down with a fever. All I could do was wait for the fever to pass and regret my stupidity.

My most fervent hope was that the innocent insects going about their daily business had somehow managed to get out of the way of the dancing flames in time.

That was last summer. We never told anyone we were the culprits. It was our guilty little secret and it was the last time Mark and I bought a whole pack of cigarettes.

We stood with our backs to the crowd sharing a cigarette and looking out over the sprawling town. The snow fell about us in the fading twilight while the anticipation of the impending sleigh rides made our hearts pound.

The whole world was white. Every house wore a thick snowy cap and pendant icicles hung from the rain gutters. Some were long and thick, dangling just inches from the ground.

"Don't ever walk under those icicles," Alan pointed out.

"Why not?"

"Because they'll kill ya."

I scanned the clusters of houses and imagined how many times I'd barely escaped death.

"Yep," said Alan, "some of those icicles probably weigh close to a ton or more. And, they're as sharp as African spears. The slightest jar, like someone slamming a door inside the house, or the echoes of a barking dog could knock one of them loose. And, if it did, that would be all she wrote. If you were standing directly underneath one of them, why before you knew it, the ice would crash right through the top of your head and push your brains right out your butt."

I shuddered at the prospect. I could picture myself stupidly walking under one of those treacherous frozen pikes just when it decided to rip loose from the awning. I saw myself stumbling about blindly in the snow, the blunt end of an icicle protruding from the top of my head and the pointy end poking out of the seat of my pants.

"Alan!" We turned to see Alan's father waving from a distance. He was pulling a tiny sled. A woolen muffler circled his throat and then with a flare, the end had been tossed over his shoulder. "Ve vill go now!"

Alan shrugged and smiled, something alive coming into his eyes. "See ya at the bottom of the hill," he said and rushed to join his father. I watched Alan and his father pull their sled to the edge of the slope.

"He's short, isn't he?" Dorothy observed. Up until now she hadn't said much, just listened intently to the conversations and kept a special eye on me.

"Heck, he's only five, what do you expect?"

"Not Alan, silly. His father. He's a tiny man."

"Yeah, I guess so."

"My mother told me if you ever want to see what a boy will look like when he grows up all you've gotta do is look at his father and that's pretty much how he'll turn out. Where's your father?"

I scanned the circles of grownups and spied my father and pointed. My dad caught my eye and waved. I waved back.

"Your father so tall and handsome, he could be a movie star."

"I think he wanted to be one once."

"He did? What happened?"

"I'm not sure. He got discouraged, I think. Mom made him give up his dream and go to college and become a lawyer instead."

"Your mother made your father give up his dream?" Dorothy asked aghast.

"A-huh. But, she said it was a just a pipe dream, so it didn't really matter."

"You're too young to understand, Benny. You're only five. When you're as old as I am," —Dorothy was seven going on thirty, "you'll realize there's no such thing as a dream that doesn't matter. When we get married, I'll never make you give up a dream. If anything, I'll encourage your dreams. I gotta go. See ya, lover."

"Don't call me lover!" I hated it when Dorothy called me "lover." I barely knew her, but apparently she had a different opinion.

Dorothy left me standing alone—which was just as well, since I had no desire to join another group. With the coming of each year, I learned to enjoy—even prefer, my moments of solitude.

There was no wind—only the stillness of the night and the falling snow. I squatted to be nearer the ground and thought I could hear the frozen flakes as they landed. No two snowflakes are the same. How could each snowflake be different? Alan had told me that, but it didn't seem possible, there were just too many of them.

A little later, my father positioned himself atop the sled and I climbed onto his back. I wrapped my arms around his neck and struggled to keep myself from spilling over with laughter as his day-old whiskers poked and tickled my wrists. In the distance, at the bottom of the hill, our two-story brick house was just a pinprick—it didn't even look real. My dad pushed off with his hands and we started down the slope— slowly at first, but gaining speed rapidly.

Fourteen blocks of continuous steep hills lay before us. Soon we were reaching speeds that I couldn't begin to calculate, but surely just as fast as a jet plane, I reasoned. The falling snow and the cold December air whipped my face and I screamed with pleasure as we sped down the center of the roadway. The sawhorses and smudge pots with wicks aflame flew past like fireflies. The blades of the sled blazed a trail of thin parallel lines and Cindy galloped alongside, panting, her short black legs kicking up snow.

That night, I lay in bed and stared at the ceiling, waiting for a car to drive by. I took great delight in the reflection of the headlights as they bounced off the snow and attached themselves to the ceiling.

They moved swiftly across the room, twisting like Jell-O when they reached the corners and slid down the wall, disappearing into the floorboards.

My radio was on and was left on all night. It kept me company. The Mills Brothers were singing about worms.

On my nightstand, the night-light glowed dimly and I could make out the painted figures on the light shade—a cowboy with his lariat suspended high over his head and his horse straining to run faster. A little dogie was trying to get away, its ears laid back like a beaten dog.

Though the sounds were muffled, I could hear my parents fighting somewhere downstairs. Not unusual. They fought a lot lately. I heard a dish crash, more angry voices, then my father's heavy footsteps climbing the stairs and heading for his study.

My mother's voice still drifted up, accompanied by the sound of more dishes breaking.

I slipped out of bed and made my way down the hall. The door to my father's study was closed; the thin lines of escaping light turned the hall jaundice. I tip-toed down the stairs hugging the banister, my mother's voice growing louder, the exploding dishes making the whole house rock, or at least it seemed that way.

"You dirty rotten son of a bitch!"

My mother was standing in the middle of the kitchen, a brown and white speckled plate in her hand. She paused a moment then threw the plate against the sink cabinet. The pieces scattered across the floor. She walked in her bare feet over the broken china to an overhead cabinet and jerked the door open so hard it slammed against another cabinet and tore loose from its hinges. I was surprised how easily the screws popped out of the wood.

She pulled more plates from the shelves then turned and looked through me as though I were invisible—and, I considered for a moment that perhaps I was. "If you want to fuck somebody, why don't you fuck me?" she shouted at the ceiling. She heaved the stack of plates at the sink and they hit like a bomb.

"I knew that shit-faced, blonde-haired, big-titted secretary was nothing but a fucking whore when I met her. You lousy cock-sucking bastard son of a bitch!"

I retreated from the doorway and ascended the stairs to my bedroom.

The door to my father's study was open now. He was sitting at his desk; his face buried in his hands.

"Is mommy going crazy?" I asked.

When he raised his head I saw something approaching sadness—or, perhaps regret in my father's eyes, but he didn't answer. I ran to my bedroom and shut the door quietly.

The room was awash in shadows and a chill drifted in from the open window. I kicked off my slippers and jumped back into bed. I pulled the covers over my head. Under my pillow, was a flashlight, a Superman comic, and six chocolate-chip cookies wrapped in Saran Wrap.

I munched the cookies while flipping the pages, reading by the beam of the flashlight. There weren't many words I could understand, but just pretending was fun, too. My mind wandered and I made up my own stories as I studied the printed words.

It was warm under the covers and the air became thinner as the temperature rose. Before long my body was perspiring heavily and my pajamas grew wet and clammy, sticking to my back. I sighed and kicked off the covers. A soft breeze swept from the window, moving the curtains and washing my body with the winter's chill. I slid from the bed and dropped to my knees before the window, resting my arms on the sill, and gazed out into the night.

That was last winter. Ever since that night there had been a different look in my mother's eyes. I thought it was pain, but I wasn't sure.

Now, it was summer and the morning sun climbed higher into the sky, making its way above the rooftops of the garages that bordered both sides of the alley. Tomorrow was the day of the family picnic at Aunt Irma's. Today my mother would be working hard in the kitchen preparing pecan and lemon meringue pies. There was really nothing for me to do except get in the way. So, leaving the gravel alley with its mammoth hollyhocks and the slow moving bumblebees, I started for Alan Borgy's house.

On my way, I passed Karen Montgomery's house. She was standing in a window framed by thick drapes. She waved and I felt my feet stop moving; Karen pleased and frightened me at the same time. She was older by four years and had no fear of trying out new things or experimenting with new ideas she'd dream up while sitting on her toy box gazing out her bedroom window.

When I was four, Karen Montgomery and I would meet behind the church on the corner and let the sun warm our butts.

"It's fun," she said, hiking her dress and slipping her panties down to her knees..

"I don't know," I said doubtfully.

Karen got on her knees in the bushes and laid her head in her arms, her butt up in the air. The sun shone through a break in the branches of a pine tree and Karen moved her rear until she could feel the sun's warmth where she wanted.

"Come on, Benny," she said, "don't be a fraidy-cat."

"I'm not a fraidy-cat," I said. Directly ahead, I thought I saw a movement behind the French windows of old Miss Crabshaw's house. "Are you sure no one can see us in here?"

"No one can see us."

I unbuckled my belt and let my jeans drop to the ground. The ground wasn't as hard as it looked, and Karen was right, the sun did feel good touching my behind even though I left my underwear on. I turned my head so I could look at Karen. Her eyes were closed and her golden hair fell over her shoulders, touching the ground. "Wanna play doctor?" she asked, her eyes still closed.

"I don't know how."

"I'll teach you," she said sitting up. "I'll be the nurse and you be the patient." I started to sit up. "No," said Karen, "you just stay how you are—you're the patient. I have to examine you."

I could feel Karen's hands touching me, her hands were warm and soft. "I have to check and see if you're okay," she said, sliding her hand into my underwear and between my legs. She fondled my penis. I wasn't sure what she was doing, but it felt pretty good, whatever it was. Karen withdrew her hand when she felt the texture of my noodle change.

"Well?" I asked.

"I think you need some medicine," she said, pulling my underwear down and exposing my behind.

"What kind of medicine? I don't need no . . . ouch!" I sat up quickly, rubbing my butt where it stung. Karen was holding a twig like a syringe.

"You needed a shot," she said.

"When's it going to be my turn?" I asked, still rubbing where it'd been poked.

Karen smiled and knelt forward, laying her head in her folded arms, her butt in the air again. At first I wasn't really sure what I was expected to do. I looked at her butt. She had a fine blonde down that covered her skin. Her rear was bubbly and round. I touched her lightly. "Why don't you take my temperature and see if I have a fever?"

"With what?" I asked, looking around. "Should I use a stick?"

"Use your finger."

I touched my finger to her butt-hole and let it slide in a little ways, then quickly pulled it out.

"Nope. No fever."

"You can't tell that fast. You have to leave it in longer." I put my finger back in a little ways and left it there.

"You have to put it in farther to get my temperature good," said Karen. "Lick your finger first," she added.

"How far should I go in?"

"As far as it will go."

I licked my finger then touched the opening to her butt hole. I pushed and let my finger slide in as far as it would go. She was right, it was warm in there. Maybe she did have a fever. "How long do I have to leave it in there?"

"Until I tell you to take it out," she said. "I'll let you know when it's time," Karen assured me.

Through the branches of the bushes I watched the robins run back and forth on the lawn. The sky was a deep blue and large white clouds drifted overhead. I was wondering how a cloud knew when it was supposed to rain and when it wasn't, and if God got mad at them if they made a mistake and did the wrong thing, when a wrinkled hand burst through the bushes and grabbed ahold of Karen's skinny arm.

"Now I've got you—you bad little girl!" It was old Miss Crabshaw.

She jerked Karen to her feet and dragged her out of the bushes onto the sidewalk. "We're going to have a nice little talk with your mother about this," she snapped. Miss Crabshaw was wearing a long-sleeve, dark blue dress, and old lady shoes—black French heels with laces that crisscrossed up the front. "And she's gong to give you the whipping of your life, you nasty little lady!"

I followed as old Miss Crabshaw led Karen away, struggling and crying. She never gave Karen a chance to pull her panties up and they became tangled around her ankles making her trip and fall, skinning her knees. Karen wailed when the pain hit her.

I hated Miss Crabshaw for humiliating my friend like that. And, every night for a week prayed passionately for her demise—"and please, oh Lord, let it be a very slow and painful death."

Now, on my way to Alan Borgey's house, I stood outside Karen's house looking up at her framed figure standing in the window I wondered if she really did get the licking of her life.

Karen disappeared from the window and reappeared at the front door. She motioned me inside—her finger to her lips, signaling me to be quiet. When I reached the top step of the porch Karen whispered, "My mom and dad went on a church outing this morning. But, I couldn't go 'cause I'm sick."

"What's the matter with you?"

"I've got chicken-pox. Come on in," she whispered. I slipped inside and Karen shut the door quietly behind me.

A large throw rug carpeted the living room and family photos in dark wooden frames adorned the walls. A rocking chair with an afghan thrown over the back sat near the hearth. It was occupied by three large cats, intertwined, content and asleep.

The house was warm and the scent of oranges lingered in the air. No lights were on. Thick drapes stopped the early morning sun from entering.

"So, what's that mean, chicken-pox?"

"Oh, it's nothing, just few red spots, that's all. See?" She showed me the little red pimples that dotted her. She was wearing her nightgown and fluffy pink slippers. "But, I feel fine though."

"Your parents left you all alone?"

Karen shook her head. "My big sister is supposed to be watching me, but she snuck out with her girlfriend a little while ago. She doesn't know that I know she's gone. Come on," she said, "let go play." Karen raced up the stairs to her bedroom and I followed.

Her room was awash in filtered sunlight; the windows concealed the outdoors with layers of chiffon curtains. High shelving displayed a variety of porcelain dolls, stuffed favorites and bric-a-brac. The bedding was still askew from last nights' sleep, the pillows still held an imprint of her head.

Karen went straight to the closet and opened the door wide. "Come on," she said excitedly.

It was a large walk-in closet with a dim light overhead. Clothes hung on poles on both sides and a dollhouse sat on the floor at the far end.

"Do you know how to play 'house'?"

I thought a moment. "Is it like playing doctor?"

"Better," Karen announced with delight.

I followed Karen inside. "This is how it works," she said, pulling a blanket from the floor that had been folded and stashed beneath a pile of dirty clothes. "You're the husband and I'm the wife." I watched as Karen climbed a stool and took the loops that had been sewn onto the blanket and folded them over the hooks that had been screwed into the molding above the door.

"You've had a tough day at the office and you come home beat. It's my job as your wife to make you feel better." She climbed down from the stool. The blanket now served as a door and the real door was left open.

"Now we can hear if anyone comes home, and still have our privacy," she said matter-of-factly, walking past me.

"Why do we need privacy?"

"Grownups always need their privacy. They lock the doors to their bedroom and they do husband and wives things."

"What are husband and wives things?"

Karen pointed to the pile of dirty clothes that were on the floor beneath the clothes rack. "That's the bed. When you come home from a hard day at the office, you have to go lay on the bed to relax."

I shrugged. "Okay." That sounded easy enough. "What do I have to do?"

"Nothing. You just have to relax. I'm the wife, see, and I have to help you relax just like my mom does my dad. I know how they do it," she said dropping her voice to a whisper. "I watched them through the keyhole." She giggled and put her fingers to her lips with delight.

"Will it hurt?" I asked, remembering the shot she gave me when they played "doctor".

"No," Karen scoffed. "If you're anything like my dad, you'll like it. But first, you have to take a shower to wash off the day."

"A real shower?" I asked.

"No, just pretend. Here," she said, throwing a sheet over the clothes bar. "You just go behind the sheet and pretend you're taking a shower."

"Okay," I acquiesced and moved towards the sheet. "But first you have to take your clothes off."

"My clothes off?"

"Certainly, you can't be expected to take a shower with your clothes on, can you?"

"Well, no," I agreed. She had a point.

"Here," she said, handing me a towel. "When you finish with your shower, wrap this around your waist and go lay on the bed. I'll be busy in the kitchen tidying up and you call me and I'll come into the bedroom and help you relax, okay?"

"Okay" I said reluctantly. Can I take my clothes off in the shower?"

"No, silly. You have to take them off before you get in the shower, or you'll get them wet."

I nodded, though I was unsure as I slipped out of my shoes, pulled off my socks and unbuttoned my shirt.

To Karen's surprise, I was wearing a Superman costume underneath. I'd taken off all my clothes, but was still fully dressed. "This will never do," she said approaching me. "You'll have to take off your costume, too."

"Now, you know my secret identity," I pouted.

"Don't worry," said Karen helping me pull off my costume. "You're secret is safe with me."

"It is?" I said, brightening.

"I'll never tell anyone."

"Thank you." I was very grateful. "And, I'll always come to your rescue if you're ever in danger." Karen smiled. "So, what do I do now?"

"Take off the rest of your clothes and go behind the sheet and finish taking your shower, but take a fast one one because we don't know when my sister will be com"ing back."

"Okay." I readily agreed and stripped. Karen watched my naked bottom scoot behind the sheet.

I pretended to shower. I scrubbed under my arms and washed my chest. I turned around in a slow circle to let the water rinse the invisible soap from my body. When I was sure I was clean, I wrapped the towel around my waist and stepped from the shower. "Do I go lay down now," I asked.

Karen was busy in one corner of the closet making dinner. She turned amidst stirring the stew to answer me. "Yes," she said, "right over there."

I skipped happily to the pretend bed and threw myself into the dirty clothes. Karen turned back to preparing dinner while my head sank into the spent aroma of Karen's dirty clothes. Odors of every facet clung to the fibers and hovered in the air just above my head. I breathed deeply; the musky bouquet was intoxicating.

She smelled good—even the parts that weren't supposed to smell good, smelled good. I pulled a handful of Karen's undergarments over my head and closed my eyes. "I'm home, dear," I called out, playing my part well.

"What are you doing?" Karen asked, pulling her clothes off my face. She was standing over me.

"Relaxing," I explained.

"I'm supposed to relax you. You're not supposed to relax yourself," she insisted."We have to be very quiet so we don't wake the kids," she said, dropping her voice and kneeling next to me. "So, whatever happens you can't cry out, okay?"

I nodded and closed my eyes. It was stuffy in the closet, warm and quiet, and I felt my body begin to sweat, but I lay very still and tried to picture what Karen was doing. Karen undid my towel and soon I felt her long hair moving over my chest. It tickled but I suppressed laughing as her hair drifted along my belly and finally stopped at my legs.

Then suddenly, something felt warm and wet. I opened my eyes and looked down. Karen had my penis in her mouth . . .

. . . The Buick swerved as my father slammed on the brakes. I was catapulted out of my seat and tossed onto the floor. My father was cussing and pounding on the horn as I picked myself up.

"Are you okay, hon?" my mother asked. I nodded and climbed back into my seat. I was fine. The other car pulled off the road and my mom rolled down her window to shout obscenities as we sped past.

Though shaken, I settled back and continued to reminisce. My heart raced as I remembered those moments in Karen's closet.

Karen's head was moving up and down, sliding her mouth over my noodle like it was a popsicle. I squirmed and liked the sensation. Then I felt my penis start to grow in size, like it was getting hard. The thought occurred to me, what if she suddenly decided to chomp down and take a bite out of it!

I gasped and jumped to my feet.

"What's the matter?" She asked surprised. Had she done something wrong? Her father never acted like that whenever her mother did it to him. "Did it hurt?"

"No," I said, reaching for my underwear and Superman costume. "But, I suddenly remembered I've gotta go. My mom's expecting me. She'll be really mad if I'm late." It was the only excuse I could think of. All I knew for sure was I had to get out of there and anything I said would do.

I dressed so fast it made me light-headed. And, when I left Karen's bedroom I was confused—unsure which way led to the stairs, or where the front door was. I remembered Karen's voice calling to me. Concerned she might have hurt me, worried I might tell her mother, afraid she might end up getting a whipping over this.

"It's okay," I assured her as I hurried down the front steps into the sunshine. I sighed with relief as the summer air filled my lungs.

"Maybe we can play house again sometime," she called.

Still dazed, I nodded and smiled and waved goodbye. Karen looked uncertain, maybe even a little frightened. I hoped my smile and wave were reassuring.

Several blocks later I spied Alan sitting on his back porch, brooding. Alan brightened when he saw me approach and stood.

"Watcha doin'?" I asked.

"Nothin'." Alan was wearing green shorts with suspenders over a yellow T-shirt. It was plain his father had recently cut his hair—it was long on top and skinned around the ears.

"Wanna play scientist?" Alan asked, as if the idea had suddenly occurred to him. I agreed. It couldn't be any worse than playing house. Alan led the way to his garage.

Alan's garage was old and dusty and smelled of pesticides. It had a dirt floor and had never been used as a carport. Instead, it was used as a storage-shed to house the chemicals needed by his mother for breeding and cross-breeding new strains of iris.

An unfinished wooden workbench sat under the only window and was the main source of light. On its shelves, systematically arranged by height, were cans, jars, and containers holding pesticides, herbicides, fertilizers and acids in the form of liquids and powders that turned the air sweet with poison. The fumes stung my eyes and made my nose burn. When I swallowed, something caustic and bitter slid down my throat.

Alan handed me a shovel and told me to dig a hole in the floor of the garage. The floor of the garage was dirt. When asked why, he explained they needed a place to mix the chemicals and a hole would work well because when we were done we could fill it in to cover our tracks. Then Alan crawled onto the workbench and began selecting chemicals to be used for his game of 'scientist'. "How you coming with that hole?" he asked.

"Pretty good," I answered. I leaned on his shovel and watched Alan pull containers off the shelf. "We aren't going to get into trouble for doing this are we?"

"Nah—nobody cares."

Benny shrugged and continued digging. I remembered the time Mark Morey had assured me of the same thing on the morning we attacked Mark's house pretending it was a covered wagon and we were the Indians. Brandishing our spears and firing our imaginary arrows in the form of rocks, we managed to break every window in his basement. We caught hell for that.

Alan's older brother, Jack, appeared in the doorway. He was gesticulating awkwardly and making unintelligible sounds punctuated with squeaks.

"What?" Alan asked visibly annoyed and putting down a can of herbicide. There were more unintelligible sounds coming from Jack.

"No!" Alan snapped with exaggerated movements for the benefit of his deaf brother.

I caught the look of disappointment in Jack's face. After he'd turned and left, heading back to the house, I asked, "Why does he talk like that?"

"What? Oh, he can't hear," said Alan, jumping off the bench and slapping his hands against his pants, making little clouds of white dust.

"So?" I stopped shoveling, sure that the hole was plenty deep enough. "Why can't he talk? What's that got to do with hearing?"

"I don't know," said Alan prying the lid off one of the cartons that had been neatly lined up according to height. "He probably can't smell things very good either."

"You think so?"

"How should I know?"

"What did he want?"

"He wanted to play with us."

I looked back at the house. Jack was standing in the kitchen, the white chiffon curtains pulled back, watching us.

"Why can't he play with us?"

"'Cause he's deaf," said Alan emphatically. "He's no fun."

I removed my eyeglass frames and pretended to clean my glasses. Alan was about to ask me why I wore frames with no glass in them, but feeling Alan's eyes watching me, I looked up smiling and said, "I can't see a thing without them."

Alan wrinkled his forehead. "How come there no glass in 'em?"

"Oh, they're there alright," I said, slipping my frames back on. "They're invisible." I wore frames for the same reason Clark Kent did—and with my perfect disguise, no one, I was quite certain, would ever suspect me of really being Superboy. Still, I wondered how the real Superman managed to wear a costume underneath his regular clothing without becoming so ungodly warm. "You don't believe me?"

Alan shook his head.

"Well, I don't believe your brother's really deaf either."

Alan looked up quizzically while emptying a box of pink powder into the freshly dug pit.

"How do you know he isn't faking?" I asked.

"'Cause my father cries about it sometimes."

"You saw your father cry?" I asked, astonished.

Alan nodded. Looking back at the house, I could see Jack still watching us through the kitchen window. Alan dumped more powder into the pit, only this time the powder had little blue flakes in it.

"Alan," I began, "I fight for truth, justice and the American way." I unbuttoned my shirt and revealed my Superman costume. I buttoned my shirt back up and let Alan think about that for a moment. Alan wasn't sure what to say, so he didn't say anything. "I could pick up this whole garage and fly us all the way to the moon."

"Why don't you, then?"

I thought for a moment. "I don't have time right now. The family picnic is tomorrow and I'll have to get home soon to help my mother bake pies. Besides, I'm playing 'scientist' right now." Alan nodded, not really buying the story. "But, not playing with your brother is not the American way."

"So?"

"So, what are you—a Nazi or something?"

"I'm not a Nazi!" Alan shouted, pouring a bottle of clear liquid the pit.

"Then how come you don't do things the American way?"

Alan tossed a can of powder into the growing vat of chemicals. It floated on the surface for a moment then appeared to melt into the earthen caldron.

It didn't dawn on Alan, nor I, that the reason there were so many chemicals in his parents garage was that his mother was a world famous horticulturist. She had bred and propagated a host of prize winning iris plants with dazzling colors and blossoms so large it was astounding.

Alan thought a moment then walked briskly to the garage door and motioned to his brother. "I play with him sometimes," he said and returned to survey the scene with his hands on his hips. "You gonna help or not?"

Soon, joined by Alan's brother, the three of us set about dumping the remaining chemicals into the pit. I took great delight in the honking noises Jack made when he got excited.

Alan imagined himself a great scientist completing an experiment that had finally, after years of hard work, proven to be a complete success—his formula for rocket fuel would soon send spaceships carrying mighty payloads to Saturn and the planets beyond.

Jack wasn't sure what the heck we were doing, but he didn't particularly care, and it didn't really make any difference—he was just overjoyed to be included in the fun for a change.

It was a warm, sunny morning, but the sun was still hidden on the far side of the garage behind a row of seeping maple trees; the interior of the garage was cool and shaded. Alan took a handle that was once attached to an old shovel and used it to stir the mixture while Jack and I satisfied ourselves that the discarded jars and boxes, scattered about the dirt floor, were indeed empty.

Unexpectedly, my "super" hearing picked up the high squeal of brakes being applied. Looking out across the flat expanse of mowed grass that was Alan's backyard, I watched the door of a white Volvo fly open and Alan's father, Adolf Borgy, step out of the car.

"Oh, shit," said Alan dropping the stick into the vat and backing away.

"What?" I asked.

Alan hadn't noticed his father. What he had noticed was the vapors rising from the chemical mixture and the bubbles that had begun to ripple and move across the surface. It appeared to be boiling.

"It's gonna blow up!" Alan screamed. "Run for your lives!" He ran for the door with Jack and I hot on his trail.

Adolf Borgy was surprised, but pleased to see Jack running past laughing and caught himself smiling at his son's obvious enjoyment. But, his smile faded and he grew concerned when he saw the look of terror on Alan's face as he stormed past.

I stopped next to the Volvo and caught my breath while Alan and Jack disappeared around the side of the house.

I pulled open my shirt to reveal my red "S", and announced, "Mr. Borgy, I think this is a job for Superboy."

"Vot iz Superboy?" Adolf inquired, amused.

"You don't know who Superboy is?" I asked, aghast. Adolf shook his head. I buttoned my shirt back up. "Well, I was going to help you, but if you don't even know who I am, you can forget it."

"Vot could you help me vith?"

"Well," I said, walking away casually, "for one thing, I could have saved your garage."

"My garage?"

"But, as it is, I guess it will just have to blow up."

"Blow up?" Adolf dropped his attaché case and raced across the lawn. Looking back, I saw Mr. Borgy with a shovel in his hands, throwing dirt as fast as his arms could lift it, into the bubbling caldron.

I could almost hear the rumbling deep in the earth where the magic chemicals had seeped into the soil. The rumbling grew louder in my imagination; the houses began to shudder as I ran with all my might for home, hopping hedges and leaping driveways with a single bound. I could hear the neighbors panicking as they emerged from their dwellings and poured into the streets, shuddering and looking around nervously, terror written on their faces. When the explosion came it would light up the sky with orange and pink and leave a mushroom trail of inky smoke . . .

. . . I nearly jumped out of my seat when the spare tire in the trunk behind me thumped against the interior wall, then fell onto the jack as the car swerved and pulled off the main road.

"There it is," came my mother's lilting voice just as the Buick hit a dip in the road and sank to a new level. The springs creaked and the car bounced in smooth cadence.

I shook off reminiscing and watched Irma's farmhouse loom into view. Yep, there it was all right, I thought.

The clouds had gleaned into a nimbus thunderhead and hovered overhead, covering most of the sky. Still, there were pockets of blue scattered sporadically and the sun managed to break through at odd moments.

The Buick backfired and belched a plume of gray smoke into the country air as it left the paved road and began its ascent. There were relatives scattered here and there. A group of cousins stopped what they were doing to watch the Buick kick up white gravel as it climbed the winding drive.

I could see Irma, followed by Harold, leave by the backdoor and stop under a tree as our car neared the crest. Though these were normally the happiest moments of my life, I couldn't help but feel the foreboding that something ominous was about to happen. Maybe it was the look on Irma's face.

Jeffery got up from the kitchen table, walked to the backdoor and peered through the wire mesh. The Buick pulled to a stop and Hank killed the engine. It was as if the whole world fell silent. Jeffery tapped the screen door open with his foot and stuffed his hands into his pockets. He stood and paused, watching with interest, then made his way down the back stairs.

Nellie rushed to his side, took his arm, and walked with him. It seemed awfully strange that she would do that, but rather than think too much about it, he tried to make out the occupants of the car.

Irma turned and watched Jeffery approach. Her heart sank. She couldn't help but relive those moments when Jeffery was only four and was staying out at Pete's, soon after the abduction. She had to see for herself what the hell was going on. Doris had called her, frantic. Joanie was coming back—Jackie had taken Jeffery away and was keeping him out at Pete's farm.

At the crest of the snow-packed hill she slowed when she saw Jeffery standing by the side of the road. He seemed to perk up, becoming more alert when he spied her car, smiling, but then she saw his shoulders fall in disappointment as she coasted to a stop. She rolled down her window.

"What are you doing standing out here, honey?" She asked.

"I'm waiting for my mommy," said four old Jeffery.

Irma hesitated then asked, "are you sure she's coming?"

"Oh, she's coming all right. Mommy needs her little boy."

Jeffery was bundled up good. His platinum hair was poking out of his wool knit cap, his deep blue eyes were innocent, expectant.

"I want to go home," he added. "Mommy's coming to get me."

"How do you know she's coming, honey?" Irma asked, trying to reason with him.

"Because I've been a good boy," said Jeffery with conviction. "And, I even cleaned my room," he said proudly with a small smile.

"It's awfully cold out here," said Irma. She watched the steam escape from his mouth when he breathed. "You want to get in the car and get warm?"

Jeffery shook his head and looked back up the road. "No," he said with a sigh.

"I'll make you a nice cup of hot soup," Irma offered.

Jeffery shook his head again without looking at her, his eyes focused on the crest of that hill. "I don't want to miss her. I'm sure she's bound to be here soon."

Irma remembered driving back to her farm that day and having to pull off the road to avoid an accident because she couldn't see through her tears. She remembered laying her head in her arms and leaning on the steering wheel to cry for a long time. Now, she watched Jeffery standing there, about to see his mother for the first time in five years. Finally.

Only Nellie stayed by him; the other cousins seemed embarrassed and kept their distance. Jeffery squinted, honing his focus, trying to make out the occupants of the car. He'd never seen the driver before, but the man kind of reminded him of Clark Gable. He didn't recognize the kid in the back seat who was opening his passenger door and climbing out. But, the woman . . . my mom opened her door, laid her sleeping daughter in the seat, and stood up. There may as well have been no one else there because neither Jeffery nor she saw anyone but each other. Their eyes locked and my mother walked to him.

Jeffery was so surprised . . . so relieved . . . so dumbfounded when he saw his mother he didn't know what to do with himself—so he did nothing. He just stood there, his hands digging deeper into his pockets. He felt Nellie's grasp tighten. Raindrops, just a few at first, began to fall, hitting his shoulders. His mother's journey, walking from the parked car to where he was standing was happening so fast . . . was taking forever. He wasn't sure whether to meet her halfway or turn and run in the other direction.

The rain fell harder and drummed on the roofs of the Buick and the house. Water gurgled down the rain gutter and spilled onto the ground. No one said a word. They just watched, some happy for them, some sorry for them, some fearing for them. Harold moved, not sure what it was he was supposed to do, but determined to do something. Irma caught his sleeve. "Let them be," she said. "Just let them be."

His mother was so close he could have reached out and touched her. Her voice whispered from his past, "Big boys don't cry, hon. Now, you just hush. Big boys don't cry. You know that." Jeffery pinched his thigh through the lining of his pocket in an effort to quell his urge to weep. A sparrow sang its morning song, oblivious to the tears that ran down his cheeks. He tasted salt in the corners of his mouth and tried to smile, but only managed slightly. He felt Nellie's grasp release, as her hand dropped away.

Suddenly, his mother was holding him, hugging him, her shoulders moving as he heard her whine softly then cry. She looked at him and wiped way his tears to ease his longing, while all of those endless hours of standing in the snow kept creeping. The pain from the cold; the pain from the disappointment when her car never made it over the crest of that snow-packed hill; the pain of the loneliness was still inside but deadened, now. Now, all he wanted was to feel his mother's body pressed agains t his.

He buried his head in his mother's neck and felt her warmth, smelled her perfume and closed his eyes. This is the way it should be—this is the way it was supposed to be. The rain fell steadily, but no one seemed to notice. It didn't seem to matter.

"*The rain is God's way of crying.*" That's what Grandpa would say. "*When He weeps the whole world knows His sadness. Raindrops are His tears. He can only be consoled by being reassured of your love. That's just the way it is, boy. After all, He's only human, right?*"

She embraced him and held him close.

Inside Jeffery, a voice sobbed, crying out silently, "don't let me go. Never let this moment end. God, if you're listening, please don't take my mommy away from me again. I'll be a good boy. I promise."

Jeffery felt his mother's lips move near his ear and he heard the beginning of the lullaby she used to sing to him. "You're a very special boy, Mister Man," she whispered. "You're the only one for me in all the land . . ."

Washington Irving Elementary School stood like an ancient castle poised on the edge of an abyss. A victim of postmodernism, it's near Gothic appearance of uneven stone masonry and asbestos filled walls destined her for demolition within the next decade. But for now, her halls were alive with the chatter of young children.

My fourth-grade classroom was on the second floor near a wide stairwell. The halls were long and the ceilings high, rewarding shrieks and squeals with pleasant echoes. The classroom doors were heavy and dark.

It was the reading hour.

"Clear your desks," Mrs. Barrett announced as she pulled a hardback chair to the front of the classroom.

There was an explosion of activity as the students put away their books, shoved pencils into desks and engaged in last minute whispers.

Mrs. Barrett laid the book in her lap and waited for silence. I folded my hands and placed them on my desk, expectant, excited.

"The book we're going to read is called, 'The Adventures of Tom Sawyer'." The class grew quiet as she cleared her throat and read the title page. "'The Adventures of Tom Sawyer'," she said, "by Mark Twain." She turned the page and began.

I relaxed, supporting my chin with my fist, and gazed out the window at the falling leaves. It was chilly on that autumn afternoon. The weather was unpredictable. The first drops of what would become a long rainstorm splashed against the glass and turned the asphalt on the street below a glistening black.

Mrs. Barrett was a good reader. Her voice was soothing, her inflection entertaining. After a few moments the students forgot they were listening and began walking through the pictures unfolding behind their eyes.

Mrs. Barrett's voice faded to a soft murmur and I let my mind wander. I remembered that morning four years ago when I stood in the rain watching my mother embrace Jeffery . . .

. . . Irma and Harold, the aunts and uncles, my cousins, and even my father, who stayed in the driver's seat, remained still—as if frozen in time. I had never seen so many tears. No one thought of raising an umbrella. Colors turned rich, clothes became damp and hair drooped. I felt the rain run down my fingers and fall to the ground.

The nimbus cloud, pushed by the winds, drifted southerly and the rains abated as quickly as they began. The sun appeared and warmed the air. There was movement; some looked around as though stirring from a deep sleep. Nellie watched a blue sky slide into view then let her gaze wander to the barn where Jeffery and his mother were strolling.

Irma nudged Harold. "Go help Henry unload the car," she said, mindful of the overwhelming weight that each bore. While others drifted slowly, as though their legs were made of lead, Irma glided swiftly to the Buick and picked me up in her arms.

"There's my favorite nephew," she said and planted a kiss on my cheek..

Harold carried the pies Joanie had baked into the house and Hank followed, cradling his daughter, Holly.

"Hey, farm girl," I said, spying Nellie as Irma lowered me to the ground.

"Hey, city slicker," Nellie returned, joining us.

"What have you two got up your sleeves today?" Irma asked, feigning suspicion.

"Steve and Philip are building a tree house in the woods," Nellie answered.

"That sounds like a lot of work," Irma observed, checking her apron strings for tautness.

"Yeah," sighed Nellie.' "That's why I thought maybe we might go fishing instead."

"Yeah," I agreed, delighted. "It's stopped raining."

"It doesn't look to me like she'll be coming back, either," Irma said, surveying the sky. "It'll probably be pretty nice for the rest of the day."

"Does Paul have to come with us?" I asked, Irma caught the whine in my voice.

"Not unless you want him to," Irma conceded. "You two seem pretty responsible to me."

"Oh, we're responsible, all right," Nellie agreed. Irma smiled. "But, we're not responsible for that dead cat out behind the barn."

"What dead cat?" Irma gasped.

"Some ol' dead cat—a white one with long hair. Deader than s door nail."

Irma thought about that. "Nope," she finally decided. "I can't recall anyone who has a white cat around here. Must be a stray."

"Let's go 'n see it," I urged. Nellie and I set off for the barn.

"What do you suppose it died from?" I asked, keeping pace with Nellie's long stride.

"Beats me," said Nellie grinning. "You just here for the day or what?"

"Nope. I talked my mom into lettin' me stay for a couple of weeks."

"Good. That's plenty of time to get into a whole mess of mischief." We both laughed.

Inside The farmhouse, hank gingerly lowered Holly into a bassinet.

"I would have had this contraption set up earlier, Hank," said Harold, insuring the folding legs were locked in place. "But I didn't think you were coming. You know, being sick and all."

"Well, I can see you're feeling fine now," Harold observed. "Can I get you something to drink? Irma worked up a whole barrel of iced tea. It's sitting out there on the porch in the shade. Or, maybe you'd like some coffee. I'm sure there's probably a pot on the stove."

Henry Olstein looked out the window and watched Joanie and her son, Jeffery, disappear into the barn. "I could do with a real stiff drink right about now," he sighed. "But, I know Irma doesn't allow hard stuff in the house."

"She doesn't allow soft stuff in the house either," Harold laughed softly. "'Course, there was a time when I used to keep a bottle of homemade wine down in the basement hidden behind a winter's cord of wood. Used to sneak down there now and then and take a nip once in a while. But I felt so damned guilty hiding it from Irma I ended up using it for medicine once then dumped the rest down the sink."

Hank grinned and faced Harold. "I can't remember anyone ever referring to it as 'medicine' before."

"Wasn't my idea," said Harold. "Doctor told me I'd better drink some. Had a kidney stone. Hurt like a son of a bitch, too. Doctor asked if I had any wine and I confessed I had a stash and he told me to go down in the basement and drink a few glasses and that would rid me of my kidney stone. Hell, I figured he knew what he was talking about, being a vet and all."

"A veteran?"

"No. A veterinarian."

"You took medical advice from an animal doctor?"

"Why not? He always seemed to know the right things to give the animals around here. And, they've all faired well. I figure there can't be that much of a difference between animals and people."

Brother strolled over, balancing a cup of coffee on a delicate saucer."

"Brother," Hank greeted him, shaking his hand. "Every time I see you, you have less hair."

Brother grinned then slid his hand over his naked pate. "I'll tell you something," he said dropping his voice. "I never did like the little bastards anyway. They were always getting in the way—sticking up where I wanted them to lay down and falling out where I wanted them to grow. I used to spend so much time primping over those little buggers I'd miss the school bus in the morning when I was a kid."

Then, turning to Harold said: "Harold, I can't stand drinking coffee out of a dainty little cup like this," he complained, taking a sip. "The stress makes my balls ache."

"I feel the same way. Stay here, I'll go get you a mug."

Brother watched Harold head for the kitchen. "When you're trying to hold on to one of these little sons of bastards," he said, nodding to the cup teetering on its thin saucer, "every muscle in your neck hurts just trying to keep it from clattering." Brother took a sip and grinned. "I love complaining, don't you?"

"Love it," Hank admitted. "Did you and Harold know each other before he married your sister?"

"Hell, Harold and I have known each other since we were both children—we met when we were both around fourteen or so, I think. We grew up together around here. 'Course, I had more hair back then."

"Harold was just telling me about the time the veterinarian told him to drink some wine to get rid of his kidney stone."

"Oh yeah, I remember—I was here," Brother said. "If that wasn't the strangest doggone day I'll put in with ya." It sure was one he'd never forget . . .

. . . Harold had been limping around for two days suffering from a burning ache that began in his lower back and worked its way around to his abdomen. Sometimes the pain got so bad he thought he would cry.

At first he figured he just had a terrible case of gas. Then he thought maybe he was constipated. But, no constipation ever felt like this. Finally, he couldn't take it any longer and called the doctor. Male pride kept him from complaining to Irma, so when she announced she was going into Hillyard for groceries, Harold breathed a sigh of relief.

"You feeling okay?" She asked, pulling on a sweater.

"Oh, yeah. I feel great," Harold lied.

Irma was standing at the screen door gazing at the barn. The sun was just peeking over the roof and the long morning shadows were stretching across the farmyard.

"I think I'll take Duke with me," she said absently, pondering something quietly. "Is that all right with you?" she asked without looking back.

"Be my guest."

"He knows I'm going some place. He's out there laying by the car door just waiting for me to step over him."

"Well, Duke's no idiot. He's a smart dog." Harold shifted in his chair and tried to find a position less painful.

Irma walked to the table and gave Harold a kiss. "You look a little piqued. Don't over do it today," she advised before letting the screen door slam behind her and heading across the gravel drive.

Harold pushed himself to a standing position and limped to the backdoor to watch Irma. Life sure was a funny thing, he thought. It seemed to him that who we are and who we fall in love with was the product of every preceding moment, no matter how trivial.

You can call it fate, or destiny, or just plain dumb luck. Harold didn't give a shit what you called it. All he knew was the moment he met Irma his world changed forever. If they'd never met, then he probably would have died young or lived a meaningless existence. Without Irma, Harold would have been incomplete.

But, one day they did meet and not long after they married. They took up residence in a small two-story house that sat on the edge of the woods just outside Hillyard.

The small town of Hillyard, whether by design or by accident, became the switching yard for three railroad companies. Trains rumbled slowly through the sleepy town and many a night after making love Harold and Irma laid awake and listened to the clanging of the train bells.

That much was pleasant, but when they were switching, the colliding couplers and drawbars made the whole house shudder. Sometimes it would be so loud Harold would bolt upright out of a dead sleep, then sink back down with a sigh as each car in a string answered like an echo.

Irma and Harold finally did buy that farmhouse they'd always wanted. It was a rundown old farm that had seen better days, but most of the structures were still salvageable. Along with the farm, they inherited a passel of abandoned farm machinery, most of which Harold restored. But, some were beyond hope and never tasted the sweet earth again.

Harold watched Irma pull out of the driveway then coast down to the main road. Duke was sitting in the back peering out the rear window smiling and drooling into the radio speaker. Harold waved and when he was sure she was well on her way, he limped to the phone and called Brother.

"Brother, you gotta come out here, right away."

"Harold, is that you?"

"Well, who the hell do you think it is?"

"Say, you sound like you're in a good mood this morning."

"I'm not feeling all that great."

"What's the problem?"

"I've got the goddamned kidney stone in my gut and it's killing' me."

"Have you seen a doctor?"

"I did. He said I should drink a couple of glasses of wine and that would probably solve my problem."

"You don't know how to drink wine, do you?"

"No, not really. I suppose anybody could just drink it, but you know Irma doesn't allow that sort of thing in the house, so I guess I don't have all that much experience with it."

"How much experience do you have?"

"Well, none really. I think as a friend you oughta come on over here and make sure I don't get sick or drunk or something."

"Put Irma on."

"She's not here, she went into town." And as Harold said that the pain got so bad he didn't even bother to say goodbye, he just hung up.

When Brother arrived at the farmhouse later that morning, Harold didn't come out to greet him as he usually did. The farm was quiet. Even the two roosters, Barney and Peabrain, had taken a break from antagonizing one another and cuddled close, napping in the shade of the tool shed.

There was no answer when he knocked. The back door had been left open as it always was on nice days. Brother peered through the screen, shading his eyes from the sun. "Harold?" The spring sang off key as he pulled the door open and stepped into the kitchen.

Light poured through the chiffon curtains over the sink. The faucets shimmered with a soft luster. The sink was clear of dishes—the white porcelain sparkled. A mountain of fruit rested in a bowl on the table. The floor had been waxed and buffed. The baseboards shined, reflecting light like black mirrors. Brother paused before letting the screen door close softly behind him. "Good God almighty," he muttered. "If this ain't the cleanest damned kitchen I ever laid eyes on, I'll put in with ya."

"Is that you?" Harold hollered.

"Last time I checked it was," Brother answered, heading for the basement door.

"I'm down here," Harold shouted.

"I kind of figured that," Brother said making his way down the wooden staircase. The wood creaked with each step and the banisters wobbled. "When you gonna fix this damned thing?" Brother asked, shaking the railing until it rattled.'

"Sooner than I want to, if you keep abusing it like that," Harold snapped.

"Staircase isn't safe. You need to put in some supports. Get rid of this fir—maybe put in some oak or maple."

"Well, thank you. I don't figure I'd have a clue of what to do if it wasn't for your sound advice."

Brother laughed, but descended the last few steps cautiously. "Just trying to be helpful."

Harold was sitting on a pile of logs holding a paper cup filled with wine. The basement was dark and cool.

"Well, I see you've already gotten started," said Brother nodding to the cup Harold was holding.

"How many glasses of that stuff have you had so far?"

"I don't rightly remember. Let me think." Harold pondered that question for a moment. "How long has it been since we spoke on the phone?"

"I don't know. A couple of hours, maybe."

"Shit. Irma'll be back any minute. What the hell am I gonna do then?"

Brother surveyed the basement. It was nearly as clean as the kitchen. "Harold, you're the only one I know who has a basement that smells good."

Harold finished the cup and poured himself another.

"Let me try some of that," said Brother, taking the cup from Harold and spilling a little in the process.

"Careful there, now. That stuff is near as valuable as gold."

Brother sniffed then made a face. "Jesus Christ," he said. "This shit is stronger than skunk shit."

Harold grinned. "Not bad, huh?"

Brother took a sip and nearly choked. "It tastes like you've got kerosene or something in there."

"Only a drop."

Brother eyed Harold then took another sip and gasped. "Holy Mother in heaven, Harold. How many glasses of this stuff have you drank?

Harold shrugged. "I don't know. Five. Six, maybe. I wasn't really keeping count," he slurred.

"Jesus Christ."

"Brother, you better watch your damned language," Harold admonished, taking his cup back. "You know Irma doesn't tolerate taking the Lord's name in vain."

"Well, Irma ain't here right now."

"Don't make no difference. She'll know."

"How the hell she gonna know?""

Harold took a big slug and shook his head. "She just will, that's all. It's like when you walk into a room and you know somebody's been there. She can just tell."

"Give me that," said Brother taking the cup away from Harold. "Now I know you've had too much. You're talking like my sister's a saint."

"She just might be."

Brother grinned while he dumped the remainder of the wine down the sink. "So, how is this kidney-stone/wine—thing supposed too work?"

"I don't rightly know for sure. All I know is I gotta pee real bad right now."

Brother held up his hand to hush Harold and listened.

"What?"

"I think I just heard a car drive up."

They both sat still. The only thing Harold could hear was his heart pounding in his ears. "Are you sure?" he whispered too loudly.

Brother shook his head. "No, I'm not sure. But, I thought I heard tires rolling over gravel."

"There's no mistaking that sound."

Then they both heard a car door slam.

"Or that sound either," Brother said, jumping to his feet.

"Oh, my Lord. Gracious sakes," Harold muttered. "What am I gonna do now?"

"You stay here; I'll go upstairs," said Brother heading for the stairs.

"No. Come back. We'll hide behind the firewood."

"Harold," said Brother, dropping his voice to a harsh whisper, "for Christ sake, get ahold of yourself."

"Watch your language," Harold whispered louder. "We're gonna catch hell for this. Goddamn it!" Harold slapped his hand over his mouth. "I can't believe I said that," he whispered through his fingers, his eyes bugged in surprise.

"Relax," Brother consoled. "Just sit down and relax. I'll handle this." Brother started up the stairs, but stopped when he heard the screen door slam. Irma's footsteps crossed the kitchen. Brother followed her movements with his eyes, tracing the ceiling supports as though the kitchen floor was transparent. He heard grocery bags being deposited on the table or a counter.

"What about me?" Harold asked desperately.

"Just sit down. You'll be fine."

"But, I've gotta pee," said Harold looking around the basement frantically.

"So pee, then."

"There's no toilet down here."

"Pee on the logs."

"I can't. They'll stink up the house when I burn them."

"Pee in the sink."

"It's too high."

"Stand on of those logs you've been sitting on."

"Good idea."

Irma was arranging quart bottles of milk on a shelf in the icebox when Brother walked into the kitchen. "I didn't expect to see your car here when I drove up," she said, peaking around the refrigerator door and smiling.

"Hi, Sis," said Brother, kissing her cheek.

"Pass me the butter, will you?" Irma asked, nodding to the grocery bag. "It's always nice to see you. You should come out here more often during the week."

"Well, I'd like to, but I've got a business to run, you know. "

"I know. So, what makes today different?"

Brother paused, stumped by the question. In his hand he held a two-pound block of butter. Irma looked at him carefully before taking it out of his hand and placing it in the freezer compartment.

"Well, it was a tough question," she said, closing the freezer door and folding the grocery bag for storage. "How's Melva?"

"Melva?"

"Your wife. Sit down, Brother, before you fall down. You're breath smells like a distillery."

Brother sat at the kitchen table. "I ain't been drinking."

'Then, I'd shop around for a new mouthwash if I were you."

"I just had a sip."

"How would you like cup of coffee ?"

"That sounds good."

Downstairs, Harold pushed and tugged, then finally rolled a heavy stump to the sink. It took some time, and Harold worked up a sweat in the process. When he stopped to catch his breath he suddenly realized he wasn't in pain anymore. Say, that wine works pretty damn good, he thought.

All the while, his bladder continued to balloon, filling rapidly. He found himself doing a kind of dance to keep from peeing himself. Finally, he climbed onto the log. Not an easy task and it took several tries before he was successful.

"I would have to be wearing these miserable pants," he complained as he struggled with the buttons where the zipper should be.

Irma poured Brother a cup stirring in a little cream and a dollop of honey.

"So, tell me," she said, placing the cup before him. "What are you and Harold doing down in the basement?"

In an effort to stall while he thought of a good answer, Brother took a quick slurp of the coffee, but it was far too hot and burned his tongue. He blew on the coffee, the cup poised before his lips. The steam swirled into his face and moistened his nostrils.

Brother considered the fact that his tongue had just been scorched so bad he probably wouldn't be able to taste anything for about a week. Then, he reconciled that it wasn't all that bad, the pain would soon pass—when a blood-curdling scream came from the basement. The scream was so urgent and so loud Brother jumped to his feet without thinking, spilling the scalding coffee into his lap.

Henry Olstein turned away from the window wearing a smile. Brother had been telling him the story in hushed undertones, as if he were telling Hank a deep dark secret.

Harold returned with a mug of coffee for Brother and took the dainty cup and saucer off his hands. "What did I miss?"

"I was just telling Hank about the kidney stone incident," Brother replied.

"Good Lord, don't remind me," Harold pleaded.

"What was the screaming all about?" Hank asked.

"Who? Him or me? I was screaming because I'd just poured a full cup of scalding hot coffee in my lap. Harold was screaming because he'd lost his balance and fallen off the log."

"Irma must have thought we'd both gone nuts," Harold added. "Fortunately, my elbows broke the fall. Landed on my funny bone, though. Now, why the hell do they call it a funny bone? That hurt near as bad as the kidney stone.

"I pissed all over myself on top of it," he continued. "By the time the stone got to my bladder it didn't hurt anymore, anyway. The wine had me pretty numb by then, so if I was in pain I sure as heck didn't know it. But, I'll tell ya," he said, looking at Hank. "I was the most surprised fella on earth when I checked my underwear and found a rock no bigger than a coffee ground in there. It's amazing to me how something so small can cause so much damn pain."

Irma had been right, Hank thought, looking at the sky. The rain clouds weren't about to return and the weather had cleared to a pleasant summer's day.

Through the window Hank watched his son, Benny, keep pace with his cousin Nellie as they trudged up the drive and headed for the barn . . .

. . . On our way to view the dead cat, we heeded Irma's request to bury the animal and stopped off at the tool shed to grab a couple of shovels.

"I've got pee, " I announced, closing the door to the tool shed. "So, pee."

I walked around the side of the shed, out of sight of the farmhouse and unzipped my pants. Nellie joined me. "Maybe since you're gonna be here for a while, you'll get a chance to meet Jeffery," she said, watching me pee.

"Yeah, maybe." I twisted my torso back and forth, making patterns on the side of the shed. The old wood turned dark when it got wet. "I didn't know he was so handsome," I observed.

"Yep," Nellie agreed. "You're mom sure knows how to make 'em, all right." She studied my face as I concentrated on the task at hand.

Benny was slender and short for his age, but well proportioned. His eyes were dark, almost black and his eyelashes were long like a girl's. Bedroom eyes, Nellie thought. His lips were full, accented by the cleft in his chin. When he smiled a dimple formed on each cheek. "You're brother's not circumcised like you," she noted. "Why'd your mom have you circumcised?"

"I guess so I'd have a pretty dick," I answered, zipping up my pants and grabbing a shovel. "I think it's a Jewish thing."

"Jeffery's dick is pretty, too." said Nellie.

"Yeah? How would you know?"

"I've seen it," Nellie said matter-of-factly. "Hell, I've seen nearly every dick around these parts."

"When'd you see Jeffery's?"

"A while back." It had only been a day or so, Nellie thought to herself. Seemed like a hundred years ago, now.

They were at Jed Murdock's abandoned farm. Jeffery was standing in the doorway of the old shack mesmerized by the dust and cobwebs and the rumors that Jed's ghost still haunted the dwelling.

"Hey, come on," said Nellie pulling Jeffery away from the open door. "I thought you said you had to piss, again."

"I do," Jeffery admitted.

"Well, the only thing you're going to do is piss your pants hangin' around here." Nellie led Jeffery away from the shack then skipped ahead. "Come on," she hollered, "the outhouse is up here a ways."

Jeffery tried to hush her, fearing her boisterous antics might somehow disturb the dead or cause some other disruption, but this only encouraged Nellie to be more obnoxious. "Jeffery has to piss!" she shouted, cupping her hands into a megaphone.

The outhouse hadn't been cleaned or renewed in over a decade. When Jeffery pried the door open, something—he wasn't sure what—but, something bolted for the toilet hole and disappeared below.

While Jeffery was considering how foolhardy it might be to dangle his penis over that black hole—where "whatever it was" might reach up and grab it or bite it off— a spider sprang from the dark and skittered across his hand.

"Kiss my ass!" he shouted, jumping back and slapping the insect way. The outhouse door slammed closed with a bang.

"Well, I'll consider it, but you'll have to ask me a lot nicer than that," Nellie chided. She was sitting on a large rock nearby chewing on a piece of grass.

"Shit," Jeffery swore, still recovering. "There's no way I'm goiin' to piss in there."

"Nah," Nellie agreed. "I wouldn't piss in there either. Hell, who know what kind of critters might be lurking about?'

"Then, why'd you bring me here?" Jeffery demanded.

"I thought that's what you wanted. Just piss on the ground," she suggested. "That's what the kids do around here."

Jeffery took stock of his surroundings, but no place looked safe.

"Whaat are you afraid of? No one will see ya."

"You will," Jeffery retorted.

Nellie sighed. "This is the country," she said, hopping down from the rock. "Everybody's seen each other piss at one time or another. It's no big deal."

"Yeah?" Jeffery considered this.

"Sure. You want to see mine?"

He already had when she was swinging over Hangman's Creek. The image of her bare bottom was still clear in his mind.

Just then, Nellie lifted her dress until the hem touched her chin. "Ta-da!" she sang.

"Nellie!" Jeffery was shocked by her lack of grace. "Why aren't you wearing any underwear?"

She let her dress fall back. "It's uncomfortable. I don't like the way it rides up the crack of my ass. Now, it's your turn." Jeffery could feel his face flush. "You're not afraid, are ya?"

"No," Jeffery insisted. He wasn't afraid—he was embarrassed. But, if he didn't show Nellie his privates she was bound to tell the others, and that could prove to be even more embarrassing. Jeffery unbuckled his belt and unsnapped his pants.

"Now, that's more like it," said Nellie with a toothy grin. "See? That wasn't so bad was it?" She turned and strolled away while Jeffery waited impatiently for his sphincter to relax. His bladder felt like it was ready to burst.

"Ever seen a mule's dick?" she asked, turning back.

"Nope. Can't say I have," Jeffery replied as urine finally began tl dribble out.

"They've got dicks as big as a fire hose," she said, watching him pee. "Practically hangs down to their knees."

Jeffery's sphincter finally relaxed into submission, and to his great relief the urine flowed out of him like a wild river.

"Steve says the whores down in Tijuana fuck mules. He says they start training the girls when they're real young 'cause it takes years before they can stretch their pussies wide enough to fit a mule's dick inside."

Jeffery rolled his eyes and shook his head. "Hell, Steve will say just about anything if he thinks there's half a chance you're going to believe him."

"Yeah, I guess so. Only, I can't figure out how the hell those whores are supposed to do it. What do they do—get underneath and wrap their legs around the mule? I suppose they'd have to lock their ankles to keep from falling. But, then what do they do? Hang upside-down?"

"I don't think the whores in Mexico fuck mules."

"Me neither. Hell, if they did, they'd have to have arms like a monkey and legs like a giraffe's. Mules have huge bellies." Nellie watched Jeffery zip up his fly.

Two days later she and I rounded the corner of the barn and were assaulted by a stench coming from the dead cat. Nellie nudged the inert body with her foot. "Yep, deader than a door nail," she declared.

I knelt for closer inspection. Its eyes were milky, clouded by death. Flies buzzed about; I tried shooing them away. Ants were running in and out of its nose, carrying micobits of decaying feline back to their nest.

"Damn, that thing stinks," Nellie complained. She scooped up the animal with her shovel and headed for the field, just steps away.

"Come on," she said over her shoulder. "Let's bury this critter before she stinks up the whole county."

We plodded into the field over mounds of clumpy dirt, and buried the animal in a shallow grave. When we were finished, we pounded the soil flat with the back of our shovels. On our way back to the barn Nellie produced a corncob pipe.

"Where'd you get that?"

"The hardware store in Hillyard," Nellie replied, striking a wooden match on the side of the barn and lighting the pipe. She took a few puffs then passed the pipe to me.

I took a puff. "This tastes good," I said, passing the pipe back. "What are we smoking?"

"Coffee grounds," Nellie said, puffing hard to keep it lit. "Butter-Nut, I think."

We sat in the weeds behind the barn and shared the pipe. The ground, though damp, had been protected from the rain by the eves. Together we watched and listened to the neighboring farms. The air was thick with the scent of alfalfa, overturned soil and fresh manure.

I sighed and laid my head back against the side of the barn. Muffled voices were seeping through the wood.

I motioned to Nellie. She squatted next to me and put her ear up to the barn and listened. "I can't understand a word they're saying," she confessed,

"Me neither." There was note of disappointment in my voice. "But, I'd like to."

A wicked smile graced Nellie's lips. "There's a ladder to the side of the barn just around the corner."

"Where does it go?"

"To a small door that opens up behind the hay stacks in the upper loft. The barn has three lofts, so it's way up by the ceiling. It gets pretty damned hot up there, but you can still hear a mouse moving around way down on the floor. Sound carries funny inside barns. Come on." Nellie made her way to the ladder and I followed.

"The door's small," she continued, "so you have to duck your head when you go inside."

"Won't they hear us?"

"Not if we're quiet. It's almost like sound floats up. It's easy to hear things coming from the barn floor even when you're way up by the ceiling, but if you're sitting down there—you can't hear much of anything except what's nearby."

Nellie led the way and we were soon scrambling up the ladder. After climbing awhile I paused to look back. I supposed a fall from this height could easily break an arm or a leg. Above me, I saw Nellie scaling the ladder at a steady clip, never pausing to rest or adjust her footing.

At first I couldn't believe my eyes. They could be playing tricks on me, I thought, or maybe I'd suddenly acquired x-ray vision like Superman. No, that wasn't it—this was real; Nellie wasn't wearing any underwear.

I hugged the ladder and stared up. I couldn't tear my eyes away from her nakedness; her bottom swelled and relaxed as she pushed past each rung. I gripped the ladder tight to keep from falling.

As we neared the top I found myself dreading an end to our journey. The climb had been effortless. I fairly floated the last sixty feet to the upper-most loft, all the while gazing up. I was stunned by her beauty and at times nearly lost my grip in admiration.

Depending on the angle, I sometimes caught a glimpse of her "thing"—it was all smooth and puffy. I began to wish the climb would go on forever and felt a stab of disappointment as Nellie opened a tiny door and disappeared inside.

I followed suit, sliding through the narrow doorway like a snake, then slithering onto the floor of the third loft. The air was warm, warmer than outside. A subtle breeze drifted up, warm air rising, carrying the voices of my mother and Jeffery. I remained on my belly, listening, then rolled over on my back.

Nellie was already seated comfortably on a bale of hay and leaning against the side of the barn. One leg was drawn up carelessly allowing her cotton dress to ride up. She was exposing herself, but didn't appear to notice, or if she knew, she sure as heck didn't care.

She was gazing out of the loft window watching the trees move with the wind and listening to the words coming from below. She sat like a boy, but from where I lay, she was anything but that.

"That's when Doris told me that Jackie had taken you away—just for the afternoon, she'd said. But, she never brought you back," my mother's voice drifted up.

As much as I wished to remain where I was because I was enjoying the view so much, I didn't want to seem obvious, so I sat up, but remained on the floor, directly across from Nellie. To my relief the view got even better, and I pretended to eavesdrop with interest while staring off in the distance, the distance being the nakedness between Nellie's legs. To sharpen my focus, I pulled out my Clark Kent glasses and put them on.

Below us Jeffery leaned against the door of a stall and watched his mother speak. After all this time he was finally with her. After all those days and weeks and months and years of waiting, it had finally happened—whether by design or by accident, it had finally happened. They were together, even if it was only for a little while. They were really together—except now, it didn't seem real—nothing did.

A breeze whispered through the barn and fanned the bales of hay. Jeffery watched a nest of baby spiders scurry down a web, then watched the perfect lips of his mother move as she spoke of a time five years in the past.

Five years seemed so long ago and so far way she could have been speaking of a time when dinosaurs still roamed the earth. She may as well have been, for though Jeffery tried to listen, some of her words just whistled past his ears like fireflies on a moonless night.

Blue-tailed swallows swept through the barn, darting in and out of the open doors in a never-ending quest for flying insects. An owl roosting near the ceiling of the first loft watched with interest, but kept silent.

"You were probably too young to remember," said Joanie, regarding Jeffery, wondering if there would be a reaction. "But, we ended up going to court over you."

"I remember."

Joanie stopped speaking and studied her son. How much did he recall? How much could a four-year-old comprehend, let alone retain?

"It was cold in the courtroom," Jeffery said, thinking back.

He remembered wishing he'd worn something warmer that morning. When he caught himself shivering and felt his teeth chatter he pulled his coat back on. Shoot, would he ever be truly warm again? It seemed like ever since that night—Jeffery pushed the memory away—not now, he warned himself, can't think about that now. Yep, he thought, slipping on imaginary blinders, he sure as heck should have worn a warmer shirt. But, he figured he had no one to blame but himself. After all, he was four wasn't he? Certainly old enough to choose his own clothes and dress himself.

The radiators hissed under the windows, emanating heat in waves like diffused light. Earlier, when the radiators were first turned on, they made a terrible racket clanging and banging. Jeffery sat with his arms folded while the grownups shouted to be heard above the noise. After a time, the judge shushed the lawyers with a wave of his hand and both sides waited patiently for the radiators to stop their complaining before the proceedings resumed.

The steam heat never seemed to get very far—at least Jeffery never felt it. He kept waiting for it to wrap its warm fingers around him or hit him in the face like sunshine on a summer's day, but it never did. The air in the courtroom was so cold it gobbled up the warmth before it reached the gallery.

The wooden bench he was sitting on was so cold the chill seeped through his trousers and bored into his skin like iron worms. Just like that night on the porch when the stars sparkled out of focus through the cold and the angels . . . no! . . . Jeffery shook his head, knocking the memories away. He wouldn't think about that night. He refused to. He felt himself slipping—his self-control waning. He would not cry—he would not cry. He dug his thumb into his thigh as hard as he could, that helped some. Think about something else, he told himself, anything else.

Jeffery left his seat, pushed past the waist-high, wooden gate and walked to the table where his mother was seated. He hadn't seen her in weeks. He didn't know if it was alright to leave his place in the gallery, didn't know if his father would pitch a fit because he had, all he knew for sure was he needed to be near his mother now.

Joanie's eyes grew wide with surprise as Jeffery entered her line of vision then fell into her arms. He hugged her for a long moment and felt her heart beating under him as a cacophony of voices exploded from behind.

"Hey, he ain't supposed to be over there at her table!" Pete said a little too loudly. "He should be sitting over here with me!"

While Pete's attorney tried to convince Pete to lower his voice, Joanie's attorney, Frank Clayton, rose to his feet. "Your honor," he said, "this is exactly what I've been talking about. Mr. Nole continues to put his welfare above that of the child's."

"I object to that erroneous statement, your Honor," Buddy Gibson, Pete's lawyer, interrupted, rising to his feet, too.

"Erroneous?" Frank repeated, glaring at his opponent.

"And I, for one, am getting sick and tired of Mr. Clayton constantly insinuating my client has something other than his son's welfare in mind."

"And, I'm getting sick and tired of you two squabbling in my courtroom," said the judge, raising his voice. "Now, sit down; both of you. The child is fine where he is right now. There are far more important issues at hand than where the child sits or stands while we decide his future."

Joanie hugged Jeffery tighter and he finally began to feel warm again. He felt his mother's lips move against his ear. "I love you, Mister Man." Jeffery fought back the tears, but it was impossible, so he let them flow until he could finally regain control. He pinched himself until the pain blocked the wheels of thought, and let his eyes wander to the windows to watch the snowfall.

"You can always tell what kind of a snowfall it's going to be by watching it fall." That's what Grandpa Worthy said. *"You see, the faster it falls, the wetter the snow. If the snow floats slowly to the ground, like its taking its sweet ol' time getting there—that snow's good for nothing but shoveling. Heck, you can sweep that kind of snow away. But, if it rushes down, like*

it just can't wait to get to the ground, well by golly, that snow's only good for two things— sleighing and making snowmen."

Outside the snow fell rapidly past the courtroom windows. The texture would be ideal for making snowmen. Perfect, in fact . . .

. . . What seemed like a lifetime ago, while he waited in vain by the side of the road for his mother, he made a snowman. Not a big one. After all, he couldn't let himself get too engrossed in a really big project—he might miss his mother's car approaching.

But, there was certainly plenty of time to build a small snowman. The snow was wet. When Jeffery packed a snowball, the moisture bled through his woolen mittens. He knew he'd regret it later, but for now, building a snowman was far more important than keeping his hands warm.

As he worked, his muffler unraveled, and a couple of times he stepped on it, tripping, as he bent over roll his snowball. As he worked, the blood rushed to his head and made his nose run. When he wiped, the moisture from his mittens left his face wet and cold.

It didn't take long. Soon, the bottom portion of the snowman was complete. It turned out bigger than he intended, but what the heck—it was smooth and solid and almost perfectly round. Not bad, he congratulated himself, not bad at all.

He made a second snowball and began rolling. Soon it was big and round and resembled fresh bread dough. As he snuffled and worked he remembered helping his mother make bread. She always told him he was a big help. And, he knew that was true, because mothers never kid or fib about things like that.

"Just wait until we pop it in the oven," his mother said, spanking the firm mound of dough and sprinkling flour on it. "Then, the house will smell like heaven."

"Is that what heaven really smells like?" Jeffery asked.

"I'm pretty sure it does." She smiled at him. His attention was focused on the bread dough as if he expected it to suddenly move or speak. "Yep," she said, "I'm pretty sure heaven smells just like it, that, and fresh ground coffee perking away on the stove."

"I like coffee."

"Me, too. Maybe we should have a cup while we're waiting for the dough to rise."

"Okay," Jeffery agreed, "with lots of milk and sugar."

Joanie slid the ball of dough into a greased bowl. "Run your fingers over it and think nice things." Jeffery did. It was warm and smooth like skin, pleasant to touch.

It had been two years since they'd boarded a bus suddenly one day and fled the Pacific Northwest for the Deep South. Why they'd left in the first place remained a mystery, and why his mother had been so terribly upset and left his Daddy behind without even saying goodbye was a puzzle, too.

He was only two back then, so the memories were sketchy at best. And though they would eventually return to Spokane one day, for the time being, Savannah was their home.

In the summer the seagulls raised a ruckus whistling with excitement as they tagged behind the steamers and tugboats. Rising high into the skies, they circled for hours then dove into the whitecaps. Sometimes the flocks grew so large they became dense living clouds, casting a shadow that turned the water gray.

Savannah was a clean Southern town wrapped up in old Southern charm with tree-lined streets and structures built around the turn of the century. She was wrought iron painted white, green window shutters nailed open, and cloudless nights with so many stars it staggered the viewer.

Savannah had grasshoppers that could jump over rooftops, their dry wings clicking in the summer heat like paper slapping the blades of a fan. She had warm rains that came without warning and dropped tadpoles on your head when you weren't looking. Sometimes acorns popped loose from the oaks and rained-down on the roof from the sudden gusts of wind; they drum-rolled down the rain gutters and landed in the narrow garden that bordered their house.

Savannah had white houses with no basements and a mother with deep auburn hair who stood on the front porch in a summer dress drinking iced coffee. The ice cubes clinked like wind chimes whenever she raised her glass.

"When I grow up I'm going to marry you and whisk you away!" Jeffery said happily. His mother smiled and joined him on the top step, smoothing her dress before she sat.

"I think maybe it's time we went back home," Joanie said, thoughtfully.

This puzzled Jeffery. "We are home."

"This isn't home. Home is where my sisters live—and your cousins. You probably don't even remember your grandpa." Jeffery shook his head. Nope. She was right there. In fact, he barely remembered his father. "You've seen snow, but you were so young you probably don't remember that either."

Snow. He'd heard of it. His mother read him stories that told of the magical white stuff. Her voice still echoed—still spoke softly in his ear as if she were near— right there next to him—her warm breath tickling. . .

. . . He finished rolling the final snowball and placed it atop the other two. There! His snowman was complete. Well, almost—as complete as it was ever going to be. There was no time to find pieces of coal for the face. But, there was his muffler—he pulled the long wool scarf from around his neck and wrapped it around the snowman. And, there was a hat—he doffed his own and placed it on the snowman's head, then stepped back to admire his handiwork.

He blew into his hands for warmth then plunged them into his pockets and shuddered. The air had warmed enough to turn the snow into slush. The falling snow became a fine misty-rain and his newly built snowman began to waste away. Miniscule holes were opening on its skin, piercing it, boring into it's frozen heart. It was dying.

Soon it would be gone, and there would be nothing left but a memory. But, it didn't make any difference. Nothing made any difference, anymore. After missing his mother for so long, nothing mattered. He stared at the snowman almost wishing its loss would replace the ache of loneliness—but, it wouldn't.

Looking around the courtroom now, he felt as if less of him remained, too. Sometimes he felt like everyday a little bit of him flaked off and became a part of the world at large. Everyday he got smaller. Until, one day, he feared there'd be nothing left of him, or not enough left to hold his body together anymore. Then, he'd just fall apart like a house of cards collapsing under its own weight. Like that snowman, he would soon cease to exist . . .

. . . The pounding of the gavel roused Jeffery from his reverie and he jumped, but his mother's gentle hand reassured him with a touch and he relaxed again.

He was sitting next to her now—at her table. Her lawyer, Frank Clayton, acknowledged his presence with quiet nods and warm smiles, but in his eyes he thought he read uncertainty.

Jeffery slid his hands into his pockets. His fingers searched until they found an object resting near the bottom. It was still there—still with him, keeping warm. He smiled slyly and caressed it with his fingers. Right now, it was his only friend, and he kept the little figurine with him all the time. He even slept with it, clutching it tight under his pillow. Ever since that night it had become his friend— ever since that night . . . no! Can't think about that! Not now. Especially, not now.

An onslaught of mind pictures—images of snow hurling from a blackened sky and a cold porch of frozen wood exposed to the night air, descended on him. Sometimes, no matter how hard he tried he couldn't shake the memory. But, today he would. Today he would not think about it. Today the judge and the court would return him to the arms of his mother and all would be right again.

Jeffery watched the judge arch his bushy eyebrows over eyes so pale they were nearly colorless. His eyelids were puffy and Jeffery imagined if he touched them they'd give, then refill like down pillows.

Their language of legal terms held little or no meaning and he soon lost interest. Thinking of other things could only be pleasant amidst a fog of strange jargon, so his mind drifted . . .

. . . Earlier, that same day, before the sun came up, Jeffery heard his father leave by the backdoor.

The house was dark. He lay in bed trying to decide if he should try to go back to sleep or get up. Light was slipping under the door and he could hear Jackie moving around in the front room. The refrigerator door opened and closed. Far away a tractor started up and his father did whatever farmers did. He knew they had to go to court that morning, but that was hours away.

Jeffery walked out into the narrow hallway and pulled the bedroom door closed behind him.

Jackie, who was lounging in an overstuffed chair, jumped in surprise and sat up straight. "Oh," she said, relieved. "Good morning, Jeffery."

She was seated, facing the window, wearing a long flannel nightgown. When she stood, the hem touched the floor, hiding her feet.

He settled onto the small circular rug at Jackie's feet and sat near a vent where warm air from the furnace pushed its way into the room and tousled his hair. Jackie tucked her legs under her and studied him with interest. So handsome, she thought. His father was good-looking, too, but Jeffery was downright pretty—like a girl.

Jackie was anything but maternal, at least that's what she thought. She didn't have children and had never been pregnant. When the doctor told her he didn't think she could ever bear children, the news didn't seem to bother her. Hell, some women just don't need or want kids.

At least that's what she'd thought until that night a couple of weeks back. Now, she didn't know what she thought. She had grown quite fond of little Jeffery, and hadn't spoken but a fistful of sentences to his father, since then.

Up until that night, Pete and she behaved like young lovers. Doing things some might even consider perverted. But, since that night, something in her had turned off. She was as frigid as the ice on the porch.

"What are you doing?" Jeffery asked, smiling up at her.

She glanced at the book in her hands, "Just reading. You're in a good mood today," she said, changing the subject.

"I get to see my mother today," Jeffery replied. His voice was full of hope and excitement.

Not for long you won't, Jackie thought. Why do little boys love their mothers so? I never did.

Growing up on a farm with Myrtle for a mother was no picnic. She'd get her and her sisters up at the crack of dawn just to begin chores—what a pain. Jackie wasn't sure she'd ever had a chance to sleep in during her entire childhood. Oh yes, there was that one time, she thought. At least the opportunity to sleep-in had been there.

She was sixteen and pretended to be sick so she wouldn't have to go to church that Sunday—she even faked throwing up in the water closet with noises and coughing. Myrtle bought it.

"I want you to stay in bed." she ordered, pulling on her coat. "The best thing if you're ill is rest. Let your body heal itself. Do you have enough covers?" She checked Jackie's bed. "A little more wouldn't hurt."

She returned shortly and threw a comforter over Jackie. "We don't need any sick youngsters in this house," she said. "There's too much work to do." She tucked the edges of the comforter under the mattress. "There's baptisms today. That means services will be a little longer than usual."

The snow stopped falling sometime during the night, but the sky hadn't cleared, which only meant it was bound to snow again later. Myrtle hoped not. The Sunday before, it snowed during services and took nearly an hour for the congregation to dig itself out.

"I see your father has the car warmed up," she said, turning away from the window. She kissed Jackie on the forehead. "Mind you stay in bed, now," Myrtle warned as she left the room.

When Jackie heard her mother clomping down the stairs she sat up and looked out the window. Her father, Arthur, was scraping ice off the windshield; exhaust puffed out the tailpipe. Joanie was brushing snow off the taillights.

Myrtle traipsed out of the house in long strides, her muffler trailing behind her. She was followed by more siblings

"Let's get a move on," Jackie heard her say. "I don't want to be late." She peered up to the second story and saw Jackie's face in the window. "There's a kettle of hot water on the stove," she hollered, cupping her mouth. "Make yourself some tea. That'll help." Jackie waved and Myrtle turned back to her brood.

Jackie lay back down and closed her eyes. She knew there was no way she was going to go back to sleep. Instead, she tried to decide how best to use her time. It was so seldom she got a chance like this, she didn't want to waste it. She was definitely going to do something— anything she couldn't normally do.

After the car chugged out of the drive, she slipped out of bed and padded barefoot downstairs to the kitchen. It was nice being alone for a change.

She made a cup of tea and strolled through the house blowing on the liquid, willing it to cool.

She turned on the radio and waited patiently for the tubes to warm up. Finally, a station managed to find its way to the rural farm and music filled the room. It was the Paul Whiteman Orchestra and Bing Crosby was mumbling melodic.

During the musical interlude, Jackie moved to the music, careful not to spill her tea. The fire in the hearth was burning low, so she tossed a fresh log on. It complained with loud pops and sizzled as the heat consumed the damp.

The house was cold. The backdoor had been opened and closed so many times, any warmth that might have settled in the house had long since escaped. Jackie rubbed her arms to get rid of the goose bumps and stood close to the fireplace.

The phone rang. Jackie frowned at the inconvenience, but left the comfort of the fire and rushed to the kitchen to answer.

"Hi, is Jackie there?"

"Pete? Is that you?" Jackie answered.

"Last time I checked it was." They both laughed.

"You never call here."

"Yeah, well," said Pete, struggling for words. "I'm callin' now."

"A-huh," Jackie responded, waiting for Pete to go on.

"Yeah, well . . . I was calling to see if you want go for a drive or something after church."

"Or, something, huh?" Jackie smiled, mimicking his drawl.

Pete laughed nervously. "Yeah."

"Well, I ain't going to church today, Pete."

"Oh." Pete hesitated, then asked. "You're not?"

"Nope. Anyway, the roads are treacherous, it snowed all night. It's still snowing some. Besides, I'm supposed to be sick."

"What's wrong with ya?"

"Nothing, stupid. There's nothing wrong with me. I said I'm *supposed* to be sick. I ain't sick."

"Well, that's good," said Pete relieved. "What are ya then?"

Jackie paused for a moment and grinned. "Listen, Pete. Are you listening real close?"

"Yep, I'm all ears."

"Everybody's gone to church, but me. I'm all alone. I'm all by myself in this great big old house and I ain't got a stitch of clothes on."

"Yeah?" There was a pause on the other end of the line. Then, as if her words had finally taken on meaning, "you don't?"

"So, if you're thinking about getting something, I suggest you get your butt over here about as fast as you can drive."

Jackie hung up and sashayed back to the hearth. She turned the radio up louder, retrieved her cup of tea and stood before the fire swaying to the music, content and excited.

By the time Pete's truck left the main road and rumbled down the access drive, the flakes had ballooned to large puff balls and visibility had shrunk to less than a mile.

Pete jumped out of the truck, slammed the door and ran up the steps. Jackie opened the door before he could knock and greeted him with a smile.

"Well, hello there, stranger," she said.

"Hell, you ain't naked," was Pete's response.

"Not yet, I ain't. You're gonna have to go around to the backdoor."

"How come?"

"Because these are hardwood floors and you can't be dripping snow all over them. Go around to the back door and take your boots off in the kitchen."

"Oh, all right," said Pete retreating back down the steps.

"And, park your dad's truck behind the tool shed just in case," she hollered. "If worse comes to worse you can sneak out the back way and take the cattle trail back up to the main road without anybody seeing you."

While Pete moved the truck, Jackie raced upstairs. In her closet she found a box that had been placed there months back. Inside was a white cotton gown she seldom wore. But this was a special occasion. It was long and flowing, the hem barely riding above the surface of the floor. When she hurried back downstairs, it fluttered behind her like wings.

She heard Pete slam the back door just as she reached the bottom landing. He was still stomping the snow from his boots and brushing off his pants when she entered the kitchen. Pete acknowledged Jackie with a nod and a shy grin, then quickly averted his eyes to concentrate on unlacing his boots.

She watched Pete as he avoided making eye contact and thought his shyness quaint, but a little unnerving. One minute he'd be talking dirty and the next he was like a little boy—quiet and unsure.

He was easy to shock, even easier to embarrass. When he became self-conscious his ears turned red and his eyes watered.

"Take your coat off and hang it on the back of the chair."

Pete obliged, pulling a brown paper bag from a pocket and setting it on the table. "I brought us something," he said, pulling a wide-mouthed jar from the bag and holding it up to the light. An amber liquid sloshed against the glass. "Whiskey."

"Where'd you get that?" "My daddy makes it."

"Isn't that against the law?"

Pete shrugged. "You want some?"

Jackie considered the prospect. "Maybe just a little," she said.

"You got something to pour it in?" he asked, looking around.

"Leave it in the jar. I don't want it smelling up a clean glass." Jackie rummaged through a drawer until she found a straw. "Come on," she said, heading for the front room. "It's cold in the kitchen."

"Yeah, I can tell," Pete smirked.

Jackie stopped before the fireplace. "How can you tell?" "You've got nobbies."

She turned her back to the hearth, shivered, and rubbed her hands together. Pete stood beside her and unscrewed the lid. "What're nobbies?" she asked, dropping the straw into the jar.

"You wanna go first?" he offered.

"No, you go ahead." Jackie watched him take a sip then gasp at the liquor's strength.

"This is powerful stuff," Pete managed in a hoarse whisper and coughed. He passed the jar to Jackie.

"What're nobbies?" she asked again.

"Oh. Well," he said, brushing his finger over her nightgown where her swollen nipples made bumps under the satin. "Those are nobbies." He grinned and pulled his hand away.

Jackie slid the straw into her mouth and took a sip. The liquid screamed down her throat. It hit her stomach like a rock, sat there, and burned. She passed the jar back to Pete as she felt her eyes tear. It took several moments for the sensation to pass.

Pete took long draws on the straw and offered the jar to Jackie. She pretended to take a sip and handed it back.

They stood next to each other listening to the radio, not saying much of anything, and passing the jar back and forth until Pete had consumed nearly all the contents. His eyes wandered over Jackie's body from time to time. Occasionally, he snickered as he took another sip.

A half-hour passed before the room was comfortably warm. Pete was a little too warm. Beads of sweat peppered his forehead and his skin tingled. He licked his lips and noticed they were numb.

Jackie held the straw in her mouth and watched Pete weave unsteadily. She handed the jar back to him without taking a swallow. He looked at its contents aghast. "Holy shit!" he slurred.

"What's the matter?"

"Hell you drank the whole goddamned thing."

"I did?"

Pete shook his head and grinned stupidly. "You silly girl. You've just gone and got yourself all drunked up." He staggered slightly then regained his equilibrium. "You just don't understand, do you?"

"Oh, I understand, all right."

"This stuff ain't just whiskey. This here is the most powerful whiskey on the face of the planet." He tottered then continued. "Shit, my old man makes whiskey so powerful just smelling it can make you drunk." He threw his head back and laughed, then caught himself as he lost his balance and staggered. The walls were moving and the floor slid dangerously underfoot like a carnival ride. "I think I better sit down."

Jackie helped him to the couch. He felt himself sink into the cushions as they rose up to meet him. The backrest gave like a pillow of warm air and his nose lost all feeling.

"You are as hell are a pretty lady, Jackie," he slurred.

"Thank you."

"But, I ain't seen you naked yet."

"You want to see me naked?"

"Yes ma'am, I sure as hell do."

"Come on," she ordered. "Stand up."

Pete tried as best he could to regain his footing, and managed to stand upright, but not without Jackie's help. With each rushing moment, time stuttered, skipping a beat, leaving unaccountable lapses in memory. Jackie took him by the hand and led the way. His legs wobbled. The first few steps were a blur, the rest were covered in a single motion as he floated upstairs.

He floundered for the bed then sighed as he collapsed on his back. He stared at the ceiling. It took too much energy to move; his body was refusing to comply anyway. He closed his eyes and thought he felt his pants being tugged down as the room began to spin.

Moments later, or was it hours?—Pete found himself behind the wheel of his father's truck rumbling over a snow packed field through a blinding snowstorm searching for the main road.

The only thing he could see clearly, though it wavered and sometimes blanked out entirely from his inebriation, was the hood ornament, a chrome plated Pegasus. He attempted to concentrate on a point somewhere between its upraised wings.

Just when he thought he was getting the hang of it, the truck lurched violently and he was thrown forward, banging his head on the steering wheel. When he looked up, to his horror, he was racing down the main road. The steering wheel was sliding under his fingers as if it had a will of its own and the truck was swerving uncontrollably.

The snowfall had escalated into a full-blown blizzard. Visibility was now less than thirty yards as the truck fishtailed and the tires spun helplessly. He could barely feel his body as it was tossed around like a rag doll and his head slammed repeatedly against the roof of the cab.

The fence line on both sides of the road dipped and rose like ocean waves, rising and falling in unison. Barbed wire and fence posts flowed crazily, sometimes disappearing altogether, lulling Pete into a sense of grinning contentment.

His eyes closed of their own accord and taking a nap suddenly seemed like a pretty good idea. When he opened them again, only as an afterthought, the truck was veering for the ditch. Frantically, he fought with the steering wheel, until he brought the truck back under control.

A rush of heat swept over him and he broke into a sweat. He pointed the head of Pegasus to the center of the road again and made up his mind he was going to keep it there. He was determined to remain conscious long enough to get home.

In an effort to remain awake, he tried singing, but eventually his mind wandered back to Jackie. He couldn't help wondering what had happened just before he found himself in the truck.

Time hiccuped and there were lapses where his memory had been wiped clean. Images, bits and pieces of the last couple hours, jumped in and out of his head like fleas on a dog. "Come on, Pete. Slam me!" He only saw flashing moments of naked pink smeared against a backdrop of white.

Snow continued to thicken and whirl around the speeding truck. In some places the ditches had filled with snow and it was impossible to tell where the road ended and the ditch began. Snowdrifts formed on the banks, and the road ahead was a wall of white that never changed.

It occurred to him that without signposts or landmarks he couldn't tell if he was headed in the right direction. Shit, he wasn't even sure where the hell he was. Nothing looked familiar, it all looked the same—a chalky haze, a bleached out world of bitter cold and heavy snowfall.

The windshield wipers did their best to wipe the glass free of snow, but it was a losing battle. Soon it piled up so heavy the wipers just stopped.

The snow hurled out of the invisible sky and collided with the windshield, completely obscuring his vision. Pete considered driving on anyway, but reasoned if he couldn't see anything, it might be a good idea to stop. He took his foot off the accelerator and let the truck coast to a stop.

"Son of a bitch," he complained, climbing out of the cab. "Goddamned snow, anyway. Fuck you!" he shouted at the snowstorm. He kicked the driver's door closed and stomped to the front of the truck where he grabbed onto the hood ornament and gazed stupidly into the eyes of Pegasus. "And, fuck you too, you horse's ass." He laughed at his own witticism and pulled himself onto the hood.

The hood was warm, too warm in places, but it was a welcome change from the bitter cold. Moving carefully, creeping on his hands and knees, Pete made his way to the windshield.

Now, what the fuck did I do with my gloves? He wondered, watching his hands turn blue. I probably left them at Jackie's. Goddamn. That's all I need—for her parents to come home and find my gloves laying around someplace when I wasn't even supposed to be there.

"Ah, who gives a shit?" he said aloud.

He was still plenty warm from the liquor and hadn't bothered to button his wool coat. But, common sense, what little he had left, told him he should get back in the cab before he froze to death.

The wipers had bonded to the frozen glass. "Kiss my fucking ass," he sighed. He worked his fingers under the rubber strip until he got a good hold and yanked. The wiper snapped free easier than he expected, and Pete lost his balance. He tumbled off the hood and landed in the road.

"Fuck me," he muttered, sitting up. Fortunately, the snowpack had cushioned his fall and he wasn't injured. "Fuck it," he said, reaching into his pocket and pulling a cigarette free. This was as good a time as any for a smoke.

Sitting in the road, leaning against the front tire for support, he inhaled the sweet smoke. "Goddamnit to hell anyway," he complained to the world at large. The truck's engine stuttered then coughed and Pete looked at the vehicle warily. "Now, what?"

He half-expected the motor to die, but it didn't. It fussed now and then with a cough or a gasp, but kept right on running. Exhaust spewed from the tailpipes and swirled around the truck while the old engine chugged and rumbled behind him drowning out other sounds.

Pete sat and fumed, puffing angrily on his cigarette. Eventually, he forgot what he was so mad about and relaxed to enjoy the smoke. He stretched his legs out, leaned hard against the tire and closed his eyes.

Above the rumble of the engine a new sound reached his ears. He squinted through the blizzard. Out of the corner of his eye, something, moving like a specter was breaking through the veil of swirling white.

The object neared; two bright, glowing eyes, floating like giant fireflies, attempted to part the moving curtain of blowing snow. Pete thought it sounded a little like a car.

Suddenly the car burst through the shifting mask of wind and snow and was upon him. The driver saw Pete's truck before it was too late and swerved hard to avoid hitting it.

A wave of moving snow kicked up by the car's tires coursed over the hood of the truck and plastered the side of Pete's face. Pete moved one of his legs out of the way just in time, but neglected to withdraw the other.

The car swept past, rolling over his foot, crushing many small bones and tearing the ligaments away from his ankle. The driver never saw Pete, only his truck.

Pete watched the black Chevy disappear into the fog of falling snow as pain raced up his leg. "Jesus Fucking Christ!" he screamed, scrambling to his feet. "Son of a fucking bitch," he cursed, limping onto his good leg.

He looked down at his crushed boot. Blood was oozing out of the eyelets and a small pool of red was spreading into the snow.

The pain was bad, but it was bearable. "It's a goddamn good thing I'm drunk," he slurred softly. Pete climbed back into the cab and released the brake. "Fuck!" he screamed, looking at the windshield. The wipers were frozen in place. Even the one he'd managed to break free earlier had frozen up again.

Pete climbed back out of the cab and limped to the front of the truck. Once more, using Pegasus for a brace, Pete pulled himself onto the hood and this time managed to work both wipers free without incident.

The job done, and moving slowly, he retreated back on all fours until he could work his knees around the hood ornament. Then lowering himself cautiously, he reached for support with his good foot.

He let the toe of his boot skip down the grating until he found the bumper. But, as he eased his weight onto the metal, the boot slipped and the rest of his body followed quickly.

For an instant he worried what more might befall his already injured right foot and braced for an explosion of pain. What he hadn't anticipated were the wings of Pegasus.

Just before his chin slammed against the hood, the tip of a wing pierced his cheek, entered his mouth and broke off a tooth.

It took a second for Pete to realize what had happened. He moved his tongue and felt the smooth metal inside his mouth. What the fuck else could go wrong? He groaned.

He tried pulling himself free, but soon realized the angle was wrong and he'd have to climb back up on the bumper to pull Pegasus's wing out of the side of his face.

"Goddamn," he attempted to say, but it was hard to talk with his tongue bouncing off a sheath of chrome. The wind had picked up and came in gusts, shaking the truck. Even through the haze of alcohol, his ears ached. It felt like someone had stuck icicles in his ear holes and left them there.

Pete braced his hands on the hood and stepped onto the bumper. Now, all he had to do was pull his head away. But, he started feeling dizzy and was afraid he might faint so he stopped and rested.

He closed his eyes and tried to clear his head, wishing he was someplace else. To his dismay, when he opened his eyes again, he was dizzy and the snow and the cold were still there.

Suddenly it dawned on him that he hadn't been dizzy after all—it was the truck—it was moving. He hadn't reset the brake. The wind was urging the truck backwards, arcing for the ditch.

This called for desperate measures. No more time to take precautions, Pete thought, as he struggled to pull himself free.

He yanked, trying to pull himself away from the hood ornament, but somehow he'd twisted his head and the entrance wound wasn't lined up with the shape of the wing anymore. It wasn't coming out.

From his awkward position, Pete watched helplessly as the rear tires left the side of the road and the truck slid into the ditch.

"Ah, shit," he fretted as he watched the truck bed sink into the snowy deep. The sea of white advanced rapidly—engulfing the truck—a creeping ameba threatening to consume him like a foreign object. He held his breath as the snow rushed up to meet him and the prospect of being buried alive became real.

Then it stopped. The truck stopped sinking. The snow stopped moving.

The nose of the truck was now pointed to the sky. The first thing anybody was bound to see when they drove by would be his butt sticking up in the air, Pete thought. That is, if anybody ever decided to drive by. It wasn't exactly the kind of weather for a Sunday outing.

Pete figured he was probably going to die there. At least the engine was still running. There was a good chance he wouldn't freeze to death, at least not until the truck ran out of gas.

The engine coughed then backfired. The exhaust pipe was plugged with snow and the fuel pump was working overtime trying to draw the fuel uphill now. It was only a matter of time before it stalled.

Pete tried to relax. Not an easy thing to do, considering the circumstances. It would take careful thought if he was ever going to get out of this mess. He had to consider a course of action.

He could jerk his head away as hard and as fast as humanly possible, not giving a shit if the wing was lined up with the entrance wound or not. That was always an option, but it would probably rip the whole side of his face off, disfiguring him for life. No good-looking girl would give him a second glance after that. So, that was no longer an option.

He could try yelling for help, but no one was likely to hear him over the gusting wind. Besides, he'd have to open his mouth wide to yell proper and that would only rip his face open more. As it was, he'd probably have a scar there for the rest of his life—the smaller the better. Hell, no one would understand what he was yelling about anyway, what with a big chunk of chrome occupying a good portion of his mouth.

His last option was just to stay like he was—his head glued to the hood ornament of his dad's truck, his ass hanging halfway out in the road, his right foot bleeding all over the bumper—and eventually freeze to death. It was too much to think about. Pete closed his eyes in defeat and allowed the alcohol to consume him.

The snow, pushed by the wind, swirled around the truck. A drift began making its way up his legs—first over his boots, then under his pant legs, climbing higher with every gust of cold air. Before long, the falling snow began to coat his back and head. The less he moved the more it accumulated. His mind wandered and he thought his body was warming. At least, he felt more comfortable and soon drifted off to sleep.

He wasn't sure if it was the sudden drop of alcohol in his blood or a sound that awakened him. The abrupt awareness of his condition prompted a groan and he considered dying. If he had to die, he prayed it would be swift.

He imagined himself already dead, hovering overhead looking down on his grave. All of the most beautiful girls in Steptoe filed past his tombstone. Some of them broke into tears and dabbed at the corners of their eyes. Others wept openly, throwing themselves prostrate over the burial mound. Behind him he thought he heard the chugging of another engine.

A car door slammed—the sound dampened by the snow. Someone was approaching. Pete could hear muffled footsteps treading behind him.

"Pete? Is that you?" Pete couldn't move very well, but he managed to move his eyes enough to recognize the good Samaritan.

"Oh, hi, Mr. Worthy." Pete swallowed hard.

"Looks like you've gotten yourself into a bit of a pickle,," Arthur observed.

"Yeah, a pickle," Pete agreed.

"Well, just hold on there, young fella. We'll see what we can do."

As Arthur made his way back to his car, Pete's truck backfired twice, and shuddered as if suffering from the effects of the cold. The engine stopped, then sighed. "Looks like we got here just in time," Arthur hollered back to Pete. "That old truck there's the only thing that's been keeping you from turning into a Popsicle."

Arthur grabbed some tools from the trunk. then opened one of the rear doors. "Come on, girls. Reckon I'm gonna need your help here."

Five of the Worthy girls, Irma, Joanie, Iris, Doris and Lois piled out of the car and followed their father to the ditch where the truck was near buried.

They slowed their pace, holding onto one another's arms, and approached tentatively, unprepared for what awaited them. Pete was huddled over the hood covered by a blanket of white snow, only tufts of his brown hair poking through. Doris caught her breath and squeezed Joanie's arm hard when she saw the amount of blood that had soaked into the snow.

Arthur paid little mind to their gasps, but smiled inwardly at their attempts to console Pete while he examined the situation more closely.

"Well now," he said finally. "What we've got to do here is to get you to a hospital, Pete."

"Yes, sir," Pete mumbled, his tongue bouncing off the chunk of metal in his mouth.

"What I don't want to do is risk pulling you off that hood decoration there. So," he said drawing closer, "I'm gonna slide this hacksaw between the hood and the ornament and cut it off."

"My Pa's gonna kill me if I ruin his ornament," Pete whined.

"Don't worry about your Pa. Besides, I'm the one that's gonna ruin it, not you. I know your Pa. If he had to make a choice between you and that hood ornament there, you'd win hands down."

"I'm not so sure about that."

"Girls, when I tell you, I want you to take ahold of Pete's legs and lift to take the pressure off them. Be careful, looks like his foot might be hurt too—is that right, Pete?"

"Yes, sir," Pete managed.

"Lift him away, real gentle like, when I tell you, then we'll carry him to the car and lay him across your laps. Then we'll take us a trip into Spokane."

"Ah, you don't have to go to all that trouble," said Pete, grateful.

"The hell I don't. I just got out of church. I figure I'm bound to feel guilty for everything I do for the next couple of days. A good deed is just what the preacher ordered."

It didn't take long. Arthur sliced through the base of Pegasus then helped the girls carry Pete to the car. Myrtle laid a blanket over her daughters' laps to protect their dresses and Arthur slid Pete in headfirst.

The car was warm and the ride made comfortable by the cushion of the Worthy girl's laps. Joanie cradled Pete's head. He tried to stay awake, but found himself drifting in and out of consciousness. Whenever he shook himself awake he found Joanie looking down at him. When their eyes met, she smiled. Pete wanted to smile back, but couldn't. The more he studied her, the prettier she became. She was an angel, an angel of mercy. She helped save his life. The gentle rocking of the car lulled him into a deep sleep. By the time the car rolled up to the emergency doors at Sacred Heart Hospital and Pete swam back to the surface of life, he was in love.

Earlier, Jackie had watched Pete take off across the field of snow in his father's truck. She left the kitchen window to refill her cup of tea.

On the table sat Pete's gloves. She swiped them off the table and stuffed them in the pocket of her coat.

Music drifted in from the family room where the radio tubes still glowed. A fire roared in the hearth. She took her tea in by the fire and sat on the rug.

Outside the snow was falling so hard she couldn't see the main road anymore. The world was white. Jackie sipped her tea contentedly, enjoyed the snow, and admired the power of nature.

That was a long time ago, Jackie reminisced. At her feet, Jeffery was sitting on the floor fiddling with the tassels of the rug, lost in thought."It's starting to snow again," Jackie said, gazing in a trance out the window . . .

. . . Even now, hours later, while Jeffery sat in the courtroom at his mother's table, it was still snowing. The bush outside the courtroom window was bare except for little red berries that clung desperately to the branches. A bird flitted from one naked branch to another pulling the fruit pebbles free and swallowing them whole.

Suddenly, the courtroom was a confusion of gasps, hand-claps, back-slapping and sobs. He was nudged from his seat and gently prodded towards the table where his father was sitting.

"Oh, Jeffery," said Joanie, pulling him back. "I'm so sorry." Jeffery didn't understand what she had to be sorry for. "I'm so . . ." she tried to stop herself from sobbing, "so, very sorry, darling."

"It's all right, Mommy. Everything's gonna be all right," he said, patting her back.

"Oh, my God. I'm so sorry," she said, her voice cracking.

Jeffery felt his own tears welling. "Don't cry, Mommy," he pleaded.

"Let's go, squirt." It was his father.

"No." Jeffery took his mother's hand and squeezed it. "I'm going home with my Mommy."

"No, you're not—you're coming home with me."

"I don't want to go home with you," Jeffery cried, shrugging his father off."

"Tell him, Joanie," said Pete.

Joanie wiped away her tears. She'd be goddamned if she was going to give that son of a bitch the satisfaction of knowing she'd been beat. She held Jeffery at arms length. "You have to go with your father, now," she said, gazing into his wet blue eyes.

"But, I don't wanna."

"I know, honey. And I don't want you to, either. But the judge says you have to."

"I don't care what the judge says," Jeffery declared defiantly.

"Well, that's just too damned bad, isn't it?" Pete growled, grabbing Jeffery's arm and pulling him away.

He watched the courtroom doors swing closed behind him, meeting his mother's eyes in that split second before the varnished slabs of mahogany blocked his view. He saw her tears through his own as he cried out to her, kicking and screaming.

Jeffery was no stranger to the ache of loneliness, but when they reached the pickup where Pete tossed him inside, he sensed his life was just about over. He feared he might not be able to last much longer and might very possibly be dead by spring.

"Now, you shut the fuck up.," Pete shouted sand slammed the door,

Jeffery's grandfather, Arthur, stepped out of the courthouse and plunged his hands into his pockets. He shrugged off the wintry air and sighed heavy. He thought he might whistle, whistling always made him feel better, and he sure as hell needed to feel better right now. That was when the dampened wail of Jeffery's desperate cries bled out of the pickup and washed up the courthouse steps.

Pete's anger mounted. He jerked the door back open, reached inside and cuffed Jeffery on the side of the head. His cries became a whimper as Pete withdrew and slammed the truck door again. When he whirled around, he was surprised to find Arthur standing close, too close, so close their noses were almost touching.

"Pete?" said Arthur calmly.

"Yeah?" Pete tried to should tough.

"The child needs to be held close, not pushed away." Arthur stepped around Pete and opened the door. "Come here, son." Jeffery fell into his grandfather's arms and wept. Pete looked around, embarrassed, punching the toe of his boot into the snow.

"I want to go home," Jeffery cried.

"I know, son. I know this is really tough on you and you feel like the whole world is crashing in around you."

Jeffery hugged his grandfather, his crying replaced with after-jobs.

"Maybe you'd like to come home with me for the night. Get a chance to gather your wits." Jeffery nodded in agreement without raising his head. Arthur lifted him out of the truck and started for his car.

"Mind telling me what's going on?" Pete asked, perturbed at Arthur's interference.

"Jeffery's spending the night with his grandpa," Arthur said, without looking back.

Pete sighed in disgust. "Don't you think it would be nice if you asked me first?"

Arthur stopped and turned. "Pete," he said, "is it all right if Jeffery spends the night with his grandpa?"

Pete hesitated before answering. "Well, shit. Yeah, I guess so."

The next day Pete telephoned to tell Arthur he was on his way over to pick Jeffery up.'

"I don't think that's a real good idea," said Arthur, cautiously.

"Oh? And, why is that?"

"Couple of reasons."

"Well, I'd like to hear 'em."

Jeffery was sitting in a child's rocker in front of the fire hugging a stuffed animal, listening to his grandfather.

"First off, there's a big storm rolling in. I figure it'll hit in about an hour or so. And, if I remember correctly, your driving and snow storms don't go together real well."

"What's the other reason?"

"You've been drinking."

"Yeah, so?"

"I couldn't in good conscience let Jeffery ride in a truck with you driving in your condition. That would be irresponsible of me, wouldn't it? Under the circumstances, I'm sure you understand."

"Yeah, well, I guess so," Pete conceded.

"Besides his grandma and I would enjoy spending a little more time with him. You don"t mind, do you?"

"No, I guess not."

"Oh, and there is one more thing," said Arthur accepting a cup of tea from Myrtle and winking at her. "I think you might have hit Jeffery a little too hard last night."

"I guess I got a right to discipline my kid."

"You're right there, Pete. I'm in total agreement. Only it seems you hit him a little *too* hard. The side of his face is all swollen up and he's got a shiner, It might be a good idea if folks didn't see him like this. It could look real bad for you—if you know what I mean."

A few moments later when Arthur returned the receiver to its cradle, he chuckled softly. He rejoined Myrtle on the couch and sipped his tea. On the floor by Jeffery's chair Myrtle had placed a cup of hot cocoa. The marshmallows had melted to a milky goo and floated on the surface. Jeffery was waiting patiently for the liquid to cool.

Jeffery was feeling better. Being with his grandparents was almost as good as being home. Though he felt stabs of loneliness now and then, their kindness was comforting and their caring welcome. Grandma Worthy was strict, but never struck out in anger. Grandpa Worthy was wise beyond measure and had the kindest, saddest eyes Jeffery had ever seen.

"If there's a snow storm coming, then I'm a monkey's uncle," said Myrtle, taking a thoughtful sip of her tea.

Arthur laughed quietly. "Can't fool your grandma, can we Jeffery?"

Jeffery shook his head and answered seriously, "nope."

"And what's this about Jeffery's face being all battered up? The child hasn't got a scratch on him."

"He doesn't?" Arthur said in mock surprise. "Jeffery don't you have a scratch on you?"

Jeffery nodded and pulled up the leg of his pajamas. "I have a bruise. Right there. I got it when I jumped off the back porch." It happened that night. That night when the sky was clear and the air was cold and the angels . . . No! He would not think about that! Not now—not ever!

Arthur got down on the floor and examined Jeffery's tiny leg. "Yep, you sure as heck do. Well, there you go," he said, rejoining Myrtle on the couch. "The kid's practically an invalid."

Jeffery never did go back to live with his father. Grandpa Worthy always managed to come up with a new excuse every time Pete called— which, after a while, wasn't that often.

Two months later, Pete's body was found in the woods, the victim of a tragic hunting accident. The Sheriff figured it probably happened when Pete tried to scale a barbed wire fence. He must have lost his balance and shot himself in the chest, though what he hoped to bag with a shotgun at that time of year was anyone's guess.

At his father's funeral, Jackie was mysteriously absent and Jeffery was dry-eyed throughout the services. He was saddened for what might have been. If only Pete had once in his life said he loved his son. If only once he had touched him with affection, told him how proud he was of his little boy, or smiled enough or hugged him when he needed it. But, he never did, and Jeffery never felt the loss of a father, only relief.

No longer would he have to live with the fear of returning to that cold, foreboding cabin Pete called home, or sit in the dark kitchen and watch his father shovel hot oatmeal into his mouth, never looking up, never meeting his eyes, never saying good morning. No longer would he have to dread walking out of the house for fear of running into his father. No tears were shed for Peter Nole that day.

Jeffery would live with his grandparents from now on—that's what the courts decided. Of course, he would have preferred to live with his mother, but he had come to accept that as unlikely event—though he was never without hope.

Suddenly, it was silent. The fragrance of old bales of hay and barn dust hung in the air. I could hear myself breathing. The voices of Jeffery and my mother had stopped. Their words were no longer drifting skyward to the uppermost loft.

I thought perhaps it was just a pause in their conversation. But, the pause lasted too long. There was a commotion coming from the farmhouse. I watched Nellie lean close to the loft window and look out. "It's Grandpa," she whispered.

"Is that bad?"

"It's hard to say. Let's go," said Nellie, heading for the door.

"Let me go first," I volunteered.

I made my descent slowly, pretending to take extra care, making Nellie wait, letting her get close without stepping on my fingers. I gazed up wistfully at Nellie's derriere, burning the image of each naked mound into my brain. Her skin was clear and looked smooth, soft to the touch. I was tempted to reach out, but knowing Nellie, I was bound to get kicked in the head for such a foolhardy act.

I stepped reluctantly off the last rung of the ladder and raced to the front of the barn, Nellie close on my heels. Joanie and Jeffery had nearly reached the end of the drive and were heading directly for the black sedan where Arthur sat, resting his butt against the fender, his arms folded.

"Holy shit," Nellie said.

"Is there going to be trouble?"

"Let's hope not."

The cousins, all except for Paul—who elected to remain in the kitchen and watch through the screen door—stopped what they were doing and congregated with their parents.

Nellie and I trotted down the drive and were within hearing distance when Jeffery and my mother stopped before Arthur. Jeffery was squeezing my mother's hand so tight her fingers had lost color.

Joanie was the first to speak. "Hi, Daddy."

"It's been a long time, Joanie."

"Yes," she admitted. "A long time."

"Didn't you miss me?" Joanie wasn't sure how to respond. "Because I know I missed you," said Arthur. Joanie made a vain attempt to smile. "As a matter of fact, I missed you so much I had to put you on my 'Shit List'."

Joanie wasn't sure what he meant. "I've got a lot of people on my 'Shit List.' Pete was on my 'Shit List'. Then he went and got himself killed. I decided to keep him on my 'Shit List' anyway."

The phone rang and Irma left the fold. The screen door banged softly behind her as she brushed past Paul and crossed the kitchen.

"I missed you, Joanie."

"I missed you, too," she said, hugging her father. She stayed in his arms for a long time, feeling his warmth, remembering his scent. "Old Spice," she whispered.

"What's that?"

"Your after-shave—Old Spice. It smells good. I missed it."

"Okay," said Arthur as they broke their embrace, but stood close now. "You talked me into it. You're no longer on my 'Shit List'."

Descending the Cascades into the lowlands, the forest grew thick and green. A constant shower fell from a permanent overcast making even the inside of our car damp. The '53 Lincoln glided smoothly to sea level where ahead of us the Olympics poked their heads above the clouds. On our right was Mt. Baker; on our left, Mt. Rainier.

I sat behind my mother and watched my father drive. "We're here!" my father announced suddenly as the tires settled onto a bridge and the tires hummed. "This is the longest floating bridge in the world!" He kept a running monologue, pointing out landmarks as they came into view. "See that? That's the Smith Tower—the tallest building west of the Mississippi!" Anyone who may have been dozing found themselves wide awake now, rubbing their eyes and trying to absorb their new surroundings. My mother leaned forward in her seat, agape.

I wasn't at all impressed—a floating bridge, a tall building. So what? There was no gully. Damn I was going to miss that place—so many adventures, so many stolen kisses, so many wondrous times exploring, playing, thinking, daydreaming—and, what about the snow in the winter and the sun in the summer? For a ten year old, it was almost too terrible to consider never seeing those things again.

Of course, I'd known this day was coming for many months. The move to the big city had been planned since last Christmas—though not quite the way it eventually turned out.

I avoided mentioning the move or saying goodbye to my friends— never preparing any of them for it. How could I? I wasn't prepared myself. Sometimes I imagined that if I just pretended it wasn't going to happen then maybe it really wouldn't. But, in the back of my mind I always new it would and found myself dreading those final days.

Nope, I never told a soul—except maybe a cousin or two.

I was fascinated with the idea that Jeffery and I shared the same mother, but the fascination ended there. Joanie had taken her father up on his offer for unmonitored and unrestricted access to her first born. I had seen Jeffery on and off since then, but the visits were sometimes short, though it was obvious Jeffery cherished those moments however few they were.

Jeffery was almost five years older than me. It was just as hard for him to relate to his little brother as it was the other way around. Still, we got along fairly well. We never fought and seldom argued. There were even times when Jeffery would behave very brotherly and let me in on some of life's little secrets.

How long had it been now? Almost five years?

It was Christmas time, Christmas day to be exact. Usually the day was spent playing with my new toys, not even bothering to change out of my pajamas until noon. My father may have been Jewish, but my mother wasn't and Christmas was a big deal at the Olstein's.

In the midst of breakfast, while my father poured over the New York Times, and Holly struggled to unbutton the blouse of her new doll, the phone rang. I ignored it as I alternated between bites of toast and examining scrambled eggs under my new microscope.

There was a change in my mother's voice after she answered the phone. She spoke softly, almost whispering, then hung up quickly and hurried back to the table. "Hurry up and eat. We're spending the rest of the day at Grandpa's," she announced.

My father sighed as he put the paper down. "What's going on?"

"Everybody's on their way to Daddy's," she said with a smile before sliding a forkful of scrambled eggs into her mouth.

"Why?" My dad disliked having his plans changed. It was Christmas Day. He had plans to do absolutely nothing for as long possible.

"It's a housewarming and we've been invited. And, because it's Christmas Day," said Joanie, refusing to let his mood influence hers. "And, I'd like to see my parents. Oh," she said as if it was an afterthought, "and, because Jeffery's going to be there." That just about said it all.

Arthur had sold his farm, and anything else he had, and bought a large white four-story house that rested high on the banks of Spirit Lake. So, as a house warming and a way to celebrate his retirement, he invited his children to his new home for the Holidays.

We arrived a little after noon. After piling out of the car, I was pleased to see Irma and Harold pull up, too. My sister, Holly, was standing by my side, she'd just turned five.

Joanie waved to her sister, Irma, then strolled with my dad through the light snow. They'd begun holding hands again like when they first met. Joanie had forgiven Hanks sexual indiscretions with his one-time secretary, and the damage to their marriage appeared to be on the mend.

"It's Irma and Harold!" Holly squealed with pleasure. She hopped up and down and clapped her hands with excitement.

We hurried to the driver's door and pulled it open. "Well, who the heck are you guys?" Harold asked. "You can't be Benny and Holly. You're much too old to be either one of them."

"A-huh. I am so Holly," said Holly, afraid that her Uncle Harold really didn't recognize her.

"Are you sure?"

"Sure I'm sure. Just ask Benny," she said.

"I never saw her before in my life," I said in mock seriousness.

"Well, if that don't beat all. Irma, do you recognize these little people?"

I ran around to the other door and helped Irma out. Harold continued to tease Holly while I took Irma's hand and strolled with her to the house. Her skin was soft and warm. I brought the back of her hand affectionately to my cheek.

Arthur's new home was a far cry from the farm. Centralized heating was managed by a natural gas furnace. There was hot and cold running water, indoor plumbing, and a two-story garage where the cousins hung out.

The families convened in the dining room where a vast array of hot and cold foods awaited them. Bowls of salads, pastas, casseroles, garnishes and warm meats sat atop four mahogany buffets that had been pushed together.

As guests helped themselves or waited in line, they were afforded a breathtaking view of Spirit Lake through large picture windows. The north wind pushed the water into choppy waves that shimmered under the winter sun.

I helped myself to the fruit punch, then noticed my little sister standing idly by. I gave her a small paper cup of punch and filled a plate for her.

"You're a good boy, Benny."

I looked up at my grandfather who was sidestepping next to me piling food on his plate. "How's life treating you?" he asked.

"Great!"

"Hmm." Arthur scooped a large portion of strawberries into a bowl. "You've got a good attitude, Benny. Keep a good sense of humor about you and you'll live a long time."

"Yeah?" I watched Grandpa pour a generous helping of fresh cream over the strawberries.

"Do you like girls, yet?"

"I've always liked girls," I replied, happy to make the truth known.

Arthur looked over his shoulder to make sure no one saw him and sprinkled a generous helping of white sugar over his strawberries.

"Aren't you a diabetic?" I asked, as if my grandfather needed reminding.

"I didn't used to be, but the doctors claim I am now," he said filling his spoon and preparing to take a bite.

"I thought you weren't supposed to eat sugar if you were a diabetic," I said.

"According to the doctors, you're not," said Arthur, taking a large mouthful. "But, they don't know everything. Beside, I love strawberries and cream. I've never asked for much in life in the way of pleasure.

But, it seems to me if you've worked as hard as I have all your life, then a bowl of strawberries isn't too much to ask for now and then." He shoveled another helping into his mouth. "I've earned it."

I saw my grandmother, Myrtle, approaching from another room. She had her eyes fixed on Arthur, but diverted them for a moment to acknowledge me.

"Grandma's coming," I whispered.

Arthur grabbed his plate and placed it over the bowl then quickly piled on more food. "Hello, dear," he said as Myrtle joined his side. "Is everything all right?"

She squeezed his arm. "Everything's fine," she said. "It's so nice having all the children here for a change."

"Yes, indeed it is," Arthur agreed, eyeing his plate.

"Hello, Benny," she said, ruffling my hair. I knew it was her way of showing affection, but I hated it when anyone messed up my hair. "Come on, honey. Harold's telling us a story about his dog, Duke. It's so good to hear his tall tales again. Come and listen to him."

"Alright, I'll be right with you." Myrtle left his side and Arthur slid the plate off his bowl and devoured the remaining strawberries. "Ah," he sighed, setting the bowl down and smacking his lips. "Now, that's what I call a real pleasure." He winked and patted me on the head before joining his wife.

"Hey, city-slicker."

I turned. It was Nellie. She was smiling and holding a paper plate of food in one hand. "Hey farm girl."

"Grandpa's gonna kill himself if he keeps eating sugar like that."

I didn't reply—not sure what to say. The idea of dying seemed pretty far- fetched, but I didn't care for the notion of being without a grandfather either.

"All the cousins are out in the garage. Jeffery's out there, too. Wanna join them?"

When we arrived at the garage, Steve and Philip were sitting in the back of an old Studebaker with the windows rolled down and their legs draped over the backrest of the front seat. They were balancing plates in their laps and eating.

Jeffery hopped down from the workbench and greeted me with a pat on the back and a shoulder squeeze. "Hey, little brother," he said loud enough for everyone to hear.

"Where's Teresa?" Nellie asked looking around.

"In the house, she had to piss," Steve answered without looking up.

"Hey everybody, it's my little brother," Jeffery announced hugging me sideways and giving my shoulder another squeeze.

"Jesus," Nellie exclaimed surveying Jeffery. "You must have grown a foot since the last time I saw you."

"So has his dick," Steve piped.

"Yeah, at least a foot," Philip agreed and chuckled.

"Yeah?" Nellie put her plate down. "What do you say Paul?"

Paul was sitting on the workbench between a miter box and a vise, concentrating on his plate of food. He nodded and shrugged. He didn't much care to join in the banter, he'd missed breakfast and felt like he was starving to death.

"Prove it," Nellie demanded. "Show us."

"Oh, he couldn't," Steve assured her in all seriousness. "It's just so damned big. Most women pass our when they see it."

"I'll take my chances," Nellie volunteered.

"Oh, no," said Philip shaking his head, "we just couldn't burden you with such a shocking sight."

Nellie strolled to the Studebaker and took a cigarette from Philip's shirt pocket. "I've been growing some myself," she said, referring to her breasts."

"Yeah, we've noticed your butt has been getting kind of big lately," Philip quipped.

"Maybe you'd like to show us how much that's grown," Jeffery added.

"Fair enough. Show us your dick first," Nellie countered

"Ladies first," Jeffery suggested.

"Oh, but I insist," said Nellie. "Age before beauty."

"Oh, he doesn't dare," Steve cautioned. "It would be irresponsible of him."

"Maybe even place your life in danger,."said Philip.

"It's that big, huh?" said Nellie.

"Bigger," Paul corroborated.

"Yeah, " I offered, "when viewed through a magnifying glass." I was trying to hold top my end off the conversation. To my relief the comment was received with hoots of laughter.

"Besides your butt is way too big for any of us to handle anyway," Steve piped.

The laughter petered out after that remark and Nellie, though she tried to conceal it, felt a little miffed. Telling a boy he had a big dick was to flatter, telling a girl she had a big butt was an insult.

"I think Nellie has a nice butt," I said without thinking.

"You do?" Nellie asked, turning to face me, surprised.

'Yeah," I said. "Sure I do."

"When did you see Nellie's butt?" Philip demanded.

Oh great, I thought, here comes her big brother to protect her. I felt my face grow hot and my ears turn red. If I'd known it was going to turn into this, I would have kept my big mouth shut. Nellie picked up on my awkwardness and came to my defense.

"Lots of times," she said.

The grins were disappearing and eyes wavered between me and Nellie. "Lots of times?" Philip asked, raising his voice.

"Philip, you can the tough guy act," said Nellie. "Nobody's impressed." She stepped closer and took my arm. "I show Benny my butt all the time. Don't I, Benny?"

I decided to play along, thankful that she'd come to my rescue. Images of following her ass up the ladder flickered. "Yeah," I said, swallowing hard. "All the time."

"You guys are so full of shit it's pitiful," said Paul, setting his plate down and borrowing a cigarette from Philip.

"So, tell us about it, Benny," Steve said, pretending interest. "Is it big and fat and flabby and have long black hairs growing out of it?"

The boys laughed hard, forcing Philip into a hacking cough. Steve pounded his back until he caught his breath. When the laughing died down, they looked at me for a reply.

I wasn't really sure what they expected. Was I supposed to come back with a really witty remark? If I was, they were shit out of luck, because witty remarks had escaped me for the moment.

I felt Nellie squeeze my arm and knew she expected something. Hell, I thought, unless Nellie was a complete idiot, she knew damned well I'd seen her ass before. Maybe she got a kick out of teasing me with the view, I wasn't sure. But I sure as heck appreciated the opportunity and wasn't about to insult her just to appease my peers. Besides, out of all my cousins, Nellie had always been my favorite.

"No," I drawled as I felt Nellie tighten her grip on my arm. "Actually, Nellie has a real nice butt. Smooth, and round and a . . ." I searched for the right word . . . "and firm."

The reaction I received was not quite what I expected. My cousins looked at me curiously and instead of countering with quips, fell silent.

Just then, Theresa burst into the garage—the door swinging wide, slamming against the garage wall. "Come quick!" she exclaimed. "It's grandpa!" She turned without another word and ran back to the house. The cousins dropped what they were doing—snuffing their cigarettes underfoot, and raced out. I lagged behind. Nellie was still holding my arm; her grip had relaxed to an affectionate squeeze.

"Thanks," she said.

I smiled. "Don't mention it."

"You know what," she said, keeping ahold of my arm, not ready to let go just yet. "I'll let you see my butt anytime you want."

"That's the best Christmas present ever."

We shared a laugh then hurried to the house.

Blair Kellogg, Steve and Theresa's father, stopped the kids at the doorway. "It is best you stay outside for now." I looked past my uncle. Irma was kneeling by her father, holding his hand. Arthur's eyes were closed, his mouth open.

"Is he dead?" Steve asked.

"I don't know. I don't think so," Blair answered, looking anxiously over his shoulder.

My dad had his ear to Arthur's heart, then sitting up alarmed, began pressing hard with both hands, pumping up and down on his chest. "Somebody call an ambulance," he ordered.

"Harold's already called one," Irma responded.

My mother and Doris began to cry softly and the fear that the great patriarch was gone descended upon the room. "Spirit Lake only has one ambulance and it's out on a call," Harold hollered from the kitchen after hanging up the phone.

"Oh my God," Irma whispered.

"Well, there's only one thing to do," said Brother..

"We'll load him into one of the cars," said Harold, finishing Brother's thought as he entered the room.

"That's right—Hank how're you doing?"

My dad was applying mouth to mouth resuscitation—between gasps he raised his head and nodded.

"Blair's got the biggest car," said Irma.

"You've got that right kiddo," Brother agreed.

Little was said after that. The men rushed forward, lifted Arthur, and headed for the door. Hank continued pumping and breathing for Arthur while Blair ran ahead and backed his suburban to the door.

Soon Blair was speeding away to the main road as a caravan of cars tore out of the driveway and followed. Within moments it was quiet again, a cloud of powdered snow fell gently back to earth, and the air grew still. Everyone was gone—only the cousins were left behind to watch the house.

I stood in the doorway and stared at the empty driveway. Beyond it, the woods and the snow-laden branches of the mighty pines drooped, as though weeping. A doe stepped gingerly out of the shadows of green and walked cautiously onto the road. She stood looking after the train of cars as if somewhere in her animal brain she was concerned and knew what was going on. I cleared my throat and the doe looked in my direction.

"Don't scare her," Nellie whispered.

I was surprised to hear Nellie's voice so close behind me.

"She's pregnant," she said. I could feel the warmth of her breath on my neck as she spoke. "At this time of the year, she's probably having a tough time finding food. Even though its winter, she has to gain weight. If she doesn't she'll lose her foal before spring."

"How do you know that?"

"I'm a farm girl. Farm girls know these things."

When we joined the other to he living room. It was plain our cousins were still coming to terms with the catastrophe.

"I hope Grandpa's going to be okay," said Philip.

"Oh, he'll be fine," Jeffery responded, not really believing it himself. The urge to cry kept creeping up on him, but he fought it bravely. "Alright," he said recovering, "let's get this place cleaned up before they all come back. Paul, help me with the buffets. Steve, you and Philip put wraps on the food and find a place in the fridge for them."

"I'll get the vacuum," Theresa volunteered.

"Come on," said Nellie, pulling my arm. "We'll do the dishes. Here, Holly, why don't you sit down at the kitchen table. We'll hand you plastic stuff and you can dry them. That'll be a big help."

Sometime later the phone rang and Jeffery answered. His face lit up as he spoke to his grandmother and when he returned the receiver to its cradle he announced, "Grandpa's going to be okay!" There was a short celebration as we all hollered, laughed and yipped. Holly jumped up and down and clapped her hands with flat palms.

"So what happened? What's wrong with him?" Philip asked.

"He went into an insulin coma. He must have snuck some sugar. He does that sometimes," Jeffery explained. I felt a pang of guilt for knowing more about it than I was willing to confess.

The house had been cleaned, the dishes were done, the buffets returned to their original positions, and the fridge neatly packed with sealed foods. Holly eventually grew tired of drying and fell asleep sitting up at the table. I carried her to the couch and covered her with an afghan.

Somebody suggested we head back to the garage and suddenly there was a mad dash for the door. Nellie and I stayed behind and listened to their whoops and hollers as they raced across the drive.

After a few moments of awkward silence Nellie spoke. "I'll bet you haven't seen the attic, have you?" she asked slyly.

"The attic?"

"It's big. You're not going to believe your eyes."

Two flights up a winding staircase and at the end of a long, dark hall was a separate stairway. It was walled on both sides and had a steep, narrow stairwell that led to a small white door near the ceiling.

Nellie led the way, which suited me just fine because it allowed me the opportunity to look up her dress. To my surprise and utter disappointment, she was wearing pink panties. Nellie opened the door and flicked on the lights; I followed her inside.

"Surprised?" she asked.

At first, I thought she was referring to her underwear, and a response was already forming, but I quickly realized she was referring to the attic.

She wasn't kidding—it was big alright. It was bigger than big—it was enormous. Hell, it was as big as the house without room dividers.

Bare light bulbs dangled from electric cords that had been strung throughout the attic and draped over nails driven into support beams. The attic was so large that even as Nellie moved ahead flicking on lights she would momentarily disappear into a black void just before another light came on.

Nellie walked back and closed the door behind him. "Come on," she said. "Wait'll you see all the junk Grandma and Grandpa have got hidden up here. Hell, they've been married more than fifty years. You would expect them to collect a little junk in all that time, but crimony—just look at all of this!"

We strolled slowly. Boxes of every size were neatly stacked in rows and labeled. Sometimes the boxes formed their own little three-walled forts. In these miniature dwellings, floor lamps or table lamps from different eras were strategically placed.

It wasn't long before we were pulled in different directions. I discovered a box of old letters and sat on the floor to examine them. They were postmarked London, England, and written in my grandfather's hand.

Nellie settled into an old, weathered rocking chair that creaked when it moved. The attic was warm. She slipped out of her sweater and relaxed. The musty scent of old books and tired clothing floated in the thin air.

The chair was comfortable. It had a high back, wide arms, and a long rock, with a seat molded to accommodate. It was old and must have been in the family for decades. The joints had swollen and shrunk so many times the apices no longer fit and and they complained with soft creaks when they rubbed . . .

. . . "Nellie?"

What a fine chair it was, too. A rocker made of birch. Arthur must have spent a good month or more just sanding it. Sometimes he'd come into the house after being in the barn, with a fine soot of sawdust covering his arms like talc.

"Nell?" Nellie felt herself being shaken awake. She'd fallen asleep in the rocker. It was warm in the attic and she'd closed her eyes for just a moment. Closed her eyes and rocked in that t old rocker.

"I think I heard a car drive up," I said, still whispering.

Nellie opened her eyes again, stretched and yawned. "My mom says you'll be moving to Seattle in the summer. Is that true?"

"Yep." I headed for the door.

"When will I get to see you again?" Nellie asked, catching up.

I shrugged. "It's six months away. We'll probably all get together again before we leave."

"I mean after you leave. What then? When will I ever get to see you after you leave?"

I shrugged again. "I don't know."

"I"m afraid you might forget about us over here."

"I won't forget."

"I'm afraid you'll forget about me."

"I couldn't forget . . ." Nellie hugged me so suddenly it caught me off guard and nearly knocked the wind out of me. Though I felt awkward at first, it was the first time I could remember feeling loved. She was nearly two years older and smelled nice, like clean clothes and pumpkin pie.

"I'm gonna miss you, city slicker," she said, still hugging me.

"I'll miss you too, country girl," I said hugging her back.

"I mean, I'm really gonna miss you," she said.

I was quiet for a moment and let myself feel her pressing against me. "I'll miss you, too," I whispered, "especially your butt." We laughed and Nellie punched me in the arm.

"Maybe we can write to each other," Nellie suggested. "Can we do that?"

"Sure, " I said. "We can do that. "

I waited at the bottom of the stairs and watched her follow me down, hoping to get a peek at her underwear. I was shocked to see she wasn't wearing any, now. "I wish you didn't have to move," she said.

"Yeah, I'm gonna miss the view."

Nellie giggled nervously. Surprised by her embarrassment, she hurried down the last few steps. "Here," she said stuffing something into my shirt pocket. "Something to remember me by."

I looked. It was her panties. They were light pink and soft. I flushed before stuffing them into my pant's pocket.

"Thanks," I said, catching up.

"Don't mention it," she replied. "To anybody," she whispered, and giggled as she descended the grand staircase ahead of me. She pushed her dress up in the back giving me one last flash of her naked butt.

I simpered. There was no reason why I couldn't spend my summers with my cousins in Spokane, even if I *was* living in Seattle. My parents might even jump at the chance to be rid of me for a few of weeks. If I had to, I could make a general nuisance of myself so they'd welcome the opportunity, but that probably wouldn't be necessary. I was convinced I could make it happen—one way or another.

Heck, Irma always had somebody staying out at her place. And, Nellie's farm was just up the road. Just up that winding dirt road with the woods on one side and plowed fields on the other. I breathed a deep sigh of contentment.

When we returned to the living room, Holly was still napping on the couch. Blair Savage Kellogg was back and was sitting in an armchair, his fingers folded over his stomach, watching the clouds lumber slowly over Spirit Lake.

To everyone's relief Grandpa Worthy lived. I figured the crisis was over and that was that, but the incident set in motion a chain of events that would change everything forever. . .

. . . Late in January, my dad called from his office. Snow still covered the ground. The radio warned another front was moving in and to expect more snow by late evening. I was delighted by the news. I took a break from studying salt crystals under my microscope to gaze out the window.

My mother was in the kitchen preparing dinner and Holly was sprawled on the living room floor, coloring. When the phone rang, Joanie quickly rinsed her hands and answered. I didn't pay much attention to the words, but there was no mistaking the change in her voice.

"Come on," she said rushing into the front room. "Your grandfather just phoned your dad. He'll be at your father's office in an hour. So, Benny go comb your hair and put your shoes on. We're going downtown."

Hank's office was on the fifth floor of the Paulson Building. I was always impressed with the high ceiling and whirling fans in the entryway. I watched my reflection in the gray and white granite floors and walls. At the far end of the lobby, the elevator with its accordion-gate slid open as we approached.

In the bigger office buildings, pretty girls in bright uniforms operated the elevators. In the Paulson Building, it was an old man wearing an undershirt creased with wrinkles and baggy tan slacks held up with suspenders. He stayed seated on a little fold-down chair, puffing on a cigar as we filed in. He took the cigar from his mouth just long enough to nod to Joanie before letting the doors close.

Grandpa Worthy was already there, talking to my father, when we walked into Henry Olstein's law office.

"I understand it won't be long before you'll be moving," he said, taking Joanie's hand and kissing her.

"Pretty soon," said Joanie, sitting.

"As soon as school lets out," said Hank. "June." It was one of the few times he spoke during the meeting.

"Good. Why don't you take Jeffery with you when you go." Arthur leaned back in his chair, folded his arms and studied his daughter.

"For how long? How long do I get to keep him?"

"Till you get tired of him. He's yours now. Hank and I already signed the papers."

Joanie went to her father and hugged him. As she broke into tears, I caught my father's eye, who seemed to be looking for a reaction, but he got none from me.

Several months later, I sat behind my mother as our yellow Lincoln glided smoothly to sea level where Cherry Street ran into 1st Avenue. "Did you see how steep that hill was?" Hank shouted with excitement. "These hills rival anything San Francisco's got—and that's a fact."

My father wore an infectious smile as he kept the running monologue alive. "What do you think of that, Jeffery?" Hank asked, beaming.

Jeffery was sitting in the backseat next to me. "Wow," he said, leaning forward, enthralled. His eyes sparkled.

Yep, I knew this day was coming for months. Now it was upon me—like a ton of bricks dropped from a high building. I don't know why Arthur had finally decided it was time Jeffery's mother and her first born should be reunited. I didn't care to listen at the time, and now that Nellie was on my mind so much, I still didn't care.

"Hey. Mom," I asked suddenly from the back seat. "Can I spend some time at Irma and Harold's this summer?"

Joanie looked sidelong at my dad. "I don't see why not. I'll call her and see what she says."

"This is Yesler Way," Hank cut in. "Pioneer Square. You watch. This city's going to grow like nobody's business."

In the summer of 1957, I turned twelve, Jeffery turned seventeen, and Holly turned eight. We, the Olstein's, settled into a four-bedroom bungalow on Capital Hill, a district in Seattle, Washington.

All in all, by the time Christmas rolled around, and looking back on it, I had to confess, if I didn't count the accident, 1957 had been one heck of a good year. For me, it was discovering I really wanted to be a writer.

This sudden awakening came late in the evening, after penning my first short story. I sat back to admire my work. While reading, I broke into goose bumps and was overwhelmed by a deep sense of well-being as I congratulated myself. Writing was almost as satisfying as masturbating.

Rather than announcing my discovery to the world, I kept it to myself. But, from then on, after everyone was asleep and the house was quiet, I slipped out of bed to sit at my desk and write.

Of course, a lot of other things were going on in the world. In 1957, the Chevrolet became the era's most popular auto, Frisbees became the year's hottest fad and Hoola-hoops came in a close second. Hank Aaron led the Milwaukee Braves to the World Series to beat the Yankees, and Wilt Chamberlain became college basketball's most celebrated hero.

The Soviet Union launched Sputnik, much to the chagrin of the United States. But, the space race was on and I did my part by turning on the TV every afternoon at four to watch the latest episode of Flash Gordon starring Buster Crabb.

"The Bridge on the River Kwai" swept the Oscars, while Rock Hudson and John Wayne remained the two top box office draws. On television, "Gunsmoke" was the number one hit. "Leave It To Beaver" debuted to join the growing list of family shows that now included "The Danny Thomas Show" and "Father Knows Best". Broadway's biggest hit, "West Side Story", brought Bernstein worldwide acclaim.

My father walked around the house carrying a copy of Kerouac's On The Road and Holly became enthralled with Dr. Seus's The Cat In The Hat.

For the first time America heard the voices of Johnny Mathis, Patsy Cline, Brenda Lee, Ricky Nelson, The Everly Brothers, Conway Twitty, Jerry Lee Lewis, Paul Anka, Buddy Holly, Jimmy Rodgers, Sam Cooke, Jackie Wilson, and Connie Francis as they debuted on AM radio. Of course, everybody already knew who Elvis Presley was and no one was a bigger fan than me, who bebopped around the house, singing into an empty toilet paper roll. Elvis's "All Shook Up" became the year's number one hit.

During Christmas break, I wrote. Writing was about the only thing that kept my mind off sex. I began work on my first novel, an African adventure called: Mocombo. The word didn't mean anything, I made it up—but it had a nice sound to it and it soon became the name for a river, a village and even a poisonous plant.

Earlier in the year, just before the accident, I returned to Colfax, Washington and stayed with Irma and Harold for the last two weeks of August. I thought about the visit often. Just hanging onto those times seemed like the only thing keeping me alive until I could return again . . .

. . . By mid-August the mercury soared into the nineties and sometimes at night it was still too warm to sleep. I stayed in the bedroom over the porch and listened to the creaking of the porch swing while struggling to drift off.

My train had pulled into Spokane after dark that evening. By the time we made it out to Colfax and the farm, had a snack of cold chicken and potato salad and visited with Irma and Harold while Duke smacked his lips under the table, I was exhausted.

I lay on my back and listened. At night in the country on Irma's farm, the frogs called for mates and the crickets did the same while I touched myself under the sheets.

In the morning I was awakened by a voice telling me to move over, and was nudged as a body crawled into bed with me. Wearing jean overalls over a white T-shirt, Nellie beamed at me as I stumbled awake.

"Okay," she said rolling onto her side and turning her back to me. "Good night." With that she snored noisily then burst out laughing.

"If Irma walks in here, she's gonna have a cow," I said.

"Don't be such a worrywart. I told Irma when I came in I was gonna come up here and wake you up."

"Like this?"

"I didn't say how I was going to do it," she confessed.

I frowned and closed my eyes.

"You can't go back to sleep."

"I'm tired," I lied.

"I have something that will wake you up," she declared. She stood up, but managed to keep her balance on a mattress that dipped and swayed like a row boat. She unbuttoned the straps of her overalls and let them drop to her ankles. She wasn't wearing any underwear—at all.

"Oh, my God," I giggled, almost afraid to look, but daring not to. "You're crazy—definitely crazy."

Nellie laughed maniacally and jumped off the bed. "Get up lazy bones," she said, pulling on her overalls and refastening her straps. "Or," she threatened before leaving the room, "that's the last time you see my ass today."

I hurried and dressed, then made my way down the carpeted staircase.

"Good morning," Irma greeted me as I walked into the kitchen.

"Well, it's about time," Nellie teased. She was seated at the table eating cold cereal out of a large white bowl.

"What have you two got up your sleeves today?" Irma asked, placing a bowl in front of me as I slid into a chair.

"I don't know," I confessed. "Something good, I hope."

"Maybe we'll go fishin'," Nellie suggested.

Irma peeked out of the kitchen window. The sky was a deep cloudless blue. "Well, it's a good day for it. That's for sure."

It was a bright, warm, and sunny day, but in the woods it was cool in the shade of the trees. The scent of pine needles, peat, and wild flowers was intoxicating as we strolled along the well-beaten path.

"So what's it like living with Jeffery?" She'd patiently held off asking until now, hoping I would eventually volunteer the information, but it didn't look like I was going to, so she nudged me.

"Kinda weird, I guess," I admitted as I quickened my stride.

"Tell me," she prompted, keeping pace.

"What's to tell?" I shrugged nonchalantly. Jesus, what's not to tell? Where do I begin? Having always been the oldest, the big brother, the first born, then suddenly relegated to the roll of the "one in the middle" was disconcerting, to say the least.

On the other hand, I quickly discovered Holly's devotion. I would always be her big brother. She didn't know Jeffery from a hole in the wall, so her support for me as the rightful heir was unwavering.

At first it seemed natural that Jeffery would want to spend so much time with his mother. After all, he'd waited most of his life to be reunited. After school, instead of watching TV, listening to records or spending time with his friends, he joined his mother in the kitchen where she was usually sitting, reading a detective novel.

I, on the other hand, preferred solitude and only ventured out of my bedroom and away from my writing desk for trips to the bathroom. Sometimes I went for a walk. On more than one occasion I returned by way of the kitchen to find Jeffery sitting on my mother's lap.

I hadn't sat on my mother's lap since kindergarten. Jeffery was nearly seventeen. Embarrassed, I muttered a weak "hi," poured myself a glass of milk and retreated quickly to my bedroom.

Later, I recalled something odd in my mother's eyes. Was it guilt? Was she embarrassed? Was she pleading for intervention? I didn't know; I couldn't tell.

I was pretty sure Jeffery's intentions were purely innocent. Maybe being without his mother's love for so many years he'd forgotten how to behave, or maybe he was just trying to catch up, picking up where he'd left off. Hell, it was just too much for me to think about.

Nellie and I sat on the edge of the pond dangling our feet in the water. On the opposite shore, a family of ducks swam about, quacking, dunking their heads into the water and ruffling their feathers, the water beading off their backs.

"Wanna swim?" Nellie asked.

"In that dirty, old pond?"

"Sure the water will be nice and warm."

"I didn't bring a swimsuit."Nellie laughed and got to her feet. "You don't wear a swimsuit in a fishing pond," she scoffed, unfastening her straps. "You have to skinny dip."

"You have to?"

She let her overalls drop to the ground and stepped out of them. "It's the law," she said pulling off her T-shirt. "An unwritten country law," she declared before jumping into the water.

The image of Nellie's nakedness right after she pulled off her T-shirt became permanently embossed on my brain. "I must be the luckiest guy in the world," I said softly.

"What?"

"Oh, nothing."

"Well, come on, then" she encouraged. Nellie paddled softly, caressing the water just below the surface.

"Is the water really warm?"

"Yes. It's like bath water."

I hesitated.

"Don't be shy," Nellie teased.

"I'm not shy," I protested.

Nellie was fourteen now and had already developed breasts—surprisingly large ones, to my way of thinking. It was hard to take my eyes off them. The only other girls I'd seen naked had been flat-chested. This was definitely a marked improvement.

"Come on," she said, smiling. "It ain't like I've never seen your dick before."

I got to me feet and nervously pulled off my shirt.

"Well, that's a start."

I slid out of my shoes and socks and pulled off my jeans and stood there in my white Penncraft underwear. Nellie was watching me. She smiled and continued to tread water. The rustling of the leaves seemed uncommonly loud—clapping facetiously.

Despite nature's applause, I tried to appear nonchalant, but I felt miserably awkward. Finally, if for no other reason than to hide my body beneath the surface of the water, I quickly stripped off my under shorts and jumped into the pond.

Nellie waited until I'd wiped the water from my eyes, then splashed more into them and laughed.

"You're kinda bashful, aren't ya?"

"No," I insisted, wiping the water from my eyes again. "Yes," I snickered, "maybe a little."

We crouched low in the water, letting the surface ride over our shoulders. It was warmer that way. I wondered if titties floated. If they did, I didn't want to miss them.

"I notice you're not wearing your Superman outfit anymore."

"I guess I kind of grew out of it."

"Do you know what you're gonna be when you grow up?

"I'm too young to grow up," I laughed.

Nellie laughed, too. "You have to sooner or later."

"No, I don't." A mosquito landed on my cheek and I smashed it with my hand, leaving a tiny dot of blood where it died. Nellie wiped the spot from my cheek.

"Mosquito guts," she said, and swiped her hand through the water.

"I ain't got time to grow up," I continued. "Besides, if I decide to stay a kid forever—up here," I said, tapping my temple, "I'll always be young. I'll never really grow old. I'll have a lot more fun than other people because I won't see things the same way."

The ducks stopped feeding for a moment and watched the two new additions to the pond with interest. "Actually," I continued, "the only thing anybody should really take seriously is their own happiness. Nothing else matters. That's what I think. Everything else is just a bunch of crap."

"You're right, It's bullish crap?" Nellie agreed.

"Yeah, fucking bullshit crap."

"Big 'ol hairy, dick fucking, bullshit crap!"

I laughed. "That's a good one."

The ducks lost interest in feeding and began a cautious journey towards their visitors. "So, what do you need to be happy?"

"I don't know," I replied, blushing.

"Oh-oh," said Nellie. "It must be something naughty."

"No, it's not," I protested, looking away. I was feeling very self-conscious. I may have thought these things, but this was the first time I'd ever confided in anyone. "Besides, you might tell."

"Tell?" Nellie was surprised. "Who the hell would I tell?" She touched my arm. "Listen, everything that happens between you and me just happens between you and me. I don't tell anybody anything. I'm my own private person. Okay?"

I nodded. "Okay."

"So, let me guess. What could it possibly be? I'll bet it has something to do with girls." I grinned. "And sex—I'll bet it has a lot to do with sex. Lot's of sex." I laughed and looked away again. "Hey, you can tell me anything, Benny. No one will ever know what we say to each other. It'll always be our secret," she confided.

I regarded her, searching her eyes. She was definitely telling the truth, I concluded.

"You can tell me anything," she repeated.

"Anything?"

"Yep." She continued to tread water, then suddenly her head slipped beneath the surface. Soon she reemerged and swiped her hair away from her eyes. "I'm back," she sang.

"Welcome back."

"I just wanted to take another look at your dick."

"You did not," I scoffed.

"Yeah," she said, seriously. "And, I can report that my suspicion was correct. It has grown in size." I felt a flush of embarrassment. "Are you going to tell on me?"

"Of course not."

"See? That's what I mean. We can say and do anything we like, as long as we keep it our secret." I nodded in agreement.

I laughed and dunked my head underwater. When I came to the surface Nellie was still grinning.

"What were you doing there?"

"Just lookin'."

"Did you see anything worth mentioning?"

"Actually," I said, growing braver, "I was checking to see if you had any hair on your pussy."

"Oh." The words were unsettling. Nellie hadn't expected me to be so blunt. "And what did you discover?"

"I didn't see any."

"That's because I shave down there."

I could see that my words had thrown her off balance. "But, first I wanted to take a gander as those tits of yours," I went on, taking full advantage of the moment. Now it was Nellie's turn to blush. "You're not bashful, are ya?" I asked, turning the tables.

"Nope," Nellie submitted. She found herself bouncing unconsciously in the water as she attempted to recover her confidence.

"Not even a little?"

Nellie was not one to allow another to get the upper hand and keep it. Half- heartedly suppressing a smile, she raised her torso out of the water until her breasts broke through the surface. They bobbed like two warm buoys, floating, riding the gentle waves. I tried prying my eyes away, but found it impossible. Now, Nellie had the advantage and she knew it. I was mesmerized.

"If you're not careful, your eyes are gonna pop out of your head and fall into the water." I pulled my eyes away to meet hers.

"I've never been this close to real boobs before," I confessed.

"Boobs? Is that what you call them?"

I nodded then let my eyes fall back to the water's surface. "I've seen pictures before, but never the real things."

Apparently the ducks hadn't either and stopped short to watch the two foreign bodies bob in the water.

Nellie swayed, pulling her breasts through the water like a paddle. She had me hypnotized; my eyes followed their every move. "You're getting sleepy," she said. "Verry sleeepy." I failed to see the humor, in fact, Nellie was doubtful I even heard her. "So, you think they're pretty then?"

"Yeah," I said. "Real pretty."

Nellie continued to move in a gentle rhythm. "You want to touch them?" She saw me nod. "Okay. Go ahead. You can if you want."

"I can?" I couldn't believe it. It would be like dipping my hands into a pirate's treasure. I reached out, and what should have been a simple task, became a major effort. The water was a thick, heavy liquid barring my way. It took forever to make contact. When I finally did, I gasped with pleasure.

Nellie, too, was enjoying the moment. Not simply because of the power she suddenly wielded, but feeling someone else touch her for the first time was enough to take her breath away. It felt nice. So absorbed were we that the sudden appearance of two people breaking through the brush caught us both by surprise. I jerked my hands away and crouched low in the water, almost fainting.

We huddled close together, Nellie behind me, as near to the bank as we could scoot. Through a veil of weeds, I watched a man and a woman, probably in their early twenties emerge from the woods. They were laughing and holding hands. They stopped briefly, spoke softly, and kissed. The man spread out a blanket for them and the woman sat down, smiling.

The ducks, too, unsure what the panic was about, but not taking any chances, hurried close to the bank, behind Nellie.

"I can't see very well," Nellie whispered, her lips nearly touching my ear. She moved close enough to rest her chin on my shoulder and peered through the shore's overgrowth. She pressed her body against mine and held onto my waist for support.

It seemed innocent enough to me at first. After all, we were cousins and she was a country girl. She was probably used to skinny dipping with her cousins and seeing them naked. At least I remembered hearing something to that effect. But, that was years ago. I doubted they still engaged in the same pastimes.

Together we watched the couple kiss and cuddle, then fall back onto the blanket. "Oh-oh," Nellie whispered. Her breath was warm and so was the water, but a chill still wormed its way up my spine. The ducks soon lost interest and began feeding while Nellie reached around and hugged my belly.

The couple on the bank were kissing passionately and the man was fiddling with the buttons on her blouse. I became aroused by Nellie's touch. I wasn't sure, but I thought I felt my penis move or throb under the water and I flushed with embarrassment. Since my back was to Nellie, I was pretty sure she hadn't noticed. If she had, she gave no indication.

Together we watched the girl's blouse fall away to reveal her breasts, plump and white in the morning sun. They kissed and fondled.

I had to look away, feeling myself grow warm. Nellie hadn't said anything, but I felt her move against my back and her hand slip lower.

Instinctively, I reached behind and pulled her close. I let my hand slip lower until I could feel her bottom. It was round and smooth and especially warm where one mound ended and the other began.

I was apprehensive, unsure whether Nellie was going to punch me in protest. But, she didn't and I was relieved. Instead, she hugged me tighter as the couple on the bank pulled off their clothes.

Time dragged, drifting; the pond's water moved lazily, swirling at a crawl, pulled by the Earth's rotation.

Far away, barely audible, voices echoed, bouncing off the valley floor.

Nellie and I were slow to react, moving in an euphoric ether as the man's head jerked up and he sprang to his feet. While he was quick to react to the sound of distant voices drawing nearer, our ears were filled with the hum of insects and paddling ducks. I watched as a look of panic spread across the woman's face and she frantically struggled to pull on her clothes.

I barely cared as the couple disappeared into the dense foliage of the woods. The fleeing figures seemed almost laughable, and their sudden retreat premature now, as the tiny voices that echoed in the distance grew fainter, and never came nearer . . .

. . . Yep, 1957 turned out to be a pretty darn good year. Almost perfect, if you didn't count the accident.

I stopped writing. I put down my pen and looked out the window. It was winter and I half-expected to see snow drifting from the sky. But, we were living in Seattle now, and it was raining. My head began to hurt again.

I popped a couple of aspirin, turned my radio on low, and lay down on the bed. The pain swelled behind my eyes and blurred my vision. Soon, my head would throb with every heartbeat. There would be no more writing for a while.

Once the headaches began, little else mattered. My main goal now was to make it through the next couple of hours. There was little I could do, but lay still and wait it out. Thinking was out of the question—not about my novel or sex or anything— not even about the accident.

The sad fact was, there was nobody to blame. I couldn't even blame myself. It would have been nice if I could have blamed somebody, even Jeffery, but it wasn't his fault either. Nope, it had just been an unfortunate accident.

It was at Jeffery's suggestion that I try out for the intramural soccer team. And, if it hadn't been for Jeffery's constant prodding I never would have. After all, I'd never been any good at sports in the past; there was no reason to believe I was suddenly going to be any good now.

Yep, if it hadn't been for my stupid brother's encouragement, I wouldn't be in the position I was in now, I thought, as the pressure continued to build inside my head. It felt like a foreign object had taken on life and was beginning to grow, pushing out in all directions, pressing my brain against the walls of my skull. Maybe I could blame Jeffery after all.

I chuckled at the thought, then grabbed my head and groaned. If I was going to blame anybody, I supposed it would have to be Aaron. But, Aaron hadn't meant to kick me in the face—it really had been an accident.

It's too bad I couldn't have made a joke about it right after it happened. I could have just stood there after my nose cartilage had been suddenly pushed against the surface of my brain as blood spurted from my face and spilled down the front of my shirt—and said: "Hey, does my head look like a ball to you?" Now, that would have been funny.

Instead, I didn't know what had hit me. I woke up on the ground staring up at the sky. The faces of my teammates hovered above me, staring down at my prone body, aghast. How humorous it would have been if I could have said in that instant, "Hey, what are looking at? Ain't cha ever seen anyone kicked in the face before?" But there was nothing funny about the pain.

My headache spread from my eyes to my temples then reached around and squeezed the back of my head. Whenever I needed to trivialize my headaches, all I had to do was remember what it felt like when I woke up on that sunny Thursday afternoon, my head on the concrete and the pain from the kick expanding inside my skull.

My teammates, wide-eyed, began pulling me to my feet. But, before my body was fully upright, I suddenly stiffened and let out a bloodcurdling scream.

I started running. I didn't know why, or where I was running to—I was just running. Running away from the blood, running away from blacking out, running away from the unspeakable pain that continued to build and swell in my brain with the intensity of a freight train. But, I hadn't run more than thirty feet before my brain shut down and I blacked out once more, my body keeling over and my head hitting the concrete with an audible crack.

Again I awoke with my teammates peering down at me. They were attempting to pull me to my feet. Again I let out a scream and started running. I made it another forty feet before the pain was too much, and I blacked out again. I kept waking up, screaming, running, and blacking out time and again.

When I was conscious, I saw my teammates running along beside me, fear struck, panic-stricken. Were they watching their teammate die? Was this the end of Benny Olstein—mild mannered kid who'd just moved to a great metropolitan city to fight a never-ending battle for truth, justice, and the American way? While I ran the color drained from the world and I tumbled into nothingness.

I don't remember waking up the fifth time. Instead, I found myself standing in the lavatory. Water was running in the sink. As I stood, dazed, trying to focus on the room, my teammates argued about whose fault it was and pulled paper towels from the rack and handed them to me.

The pain tore through my face and I screamed—long and hard. My teammates froze and stared at me in horror, then scurried about like ants whose nest had been disturbed.

At least I hadn't blacked out again. Every so often, I let out a scream, though the pain was beginning to subside. I caught the reflection of myself in the mirror. The front of my shirt was drenched. I'd never seen so much blood in my life. My face was smeared crimson. While blood dribbled from one nostril, a red stream shot out of the other, pulsating—squirting in time to my heartbeat. I screamed and passed out.

When I came to, I was weaker than I'd ever felt. I was tired of screaming and fainting and bleeding. Weariness set in—I just wanted to go home and sleep. A few teammates were still consoling me when the school principal burst into the Boy's Room.

Mrs. Bernstein had run all the way from her office on the second floor, where she first heard the ruckus. Her brown tweed skirt constricted her movements as she sprinted down the hallway, and rushed down the stairs, taking them two at a time. Now, standing in the doorway where all the noise was coming from, she was aghast. Paper towels smeared with blood were strewn from one end of the lavatory to the other.

She put her hands on her hips. "What the hell is going..." she recognized me and without another word came to my side. She dampened some paper towels in cold water and dabbed my face cautiously, cleaning me up as best she could. "What happened?"

"Benny got kicked in the face," someone said and I broke down and cried.

"Come on," she said, putting her arm around me and holding a wet towel against my face. I think the bleeding will stop pretty soon. I'll take you home."

She was right. The bleeding did seem to stop. But, for the rest of the day and that night, my sinuses felt like they were draining. So, I swallowed. And I continued to swallow even while I slept. In the morning I felt weaker still.

I stumbled into the front room and sat at the dining room table, dazed. My father was hurrying about, grabbing a couple of pens and legal pads and stuffing them into his briefcase. "How're you feeling?" he asked.

I felt lousy. I was nauseous and my head was spinning like I had the flu, but at least the pain was gone. "Not so good," I replied. "I don't think I should go to school today."

"No," Hank agreed, closing his briefcase and donning his gray homburg. "You probably shouldn't. Not after what happened." He patted me on the back. "Big case today."

"Oh?"

"Yep. A murder case. I love murder cases. How do I look?"

I attempted to appraise my father. "Good."

"Wish me luck," he said, smiling.

"Good . . ." was all I managed before my stomach gave up its loathsome burden. It all came up at once. There was no dry heaving. Just one massive regurgitation of semi-clotted blood that made a puddle the size of a wading pool on the carpet at my feet.

Jeffery stepped from the kitchen and stood in the doorway. "Holy shit," he said.

Hank had made it all the way to the door, but closed it slowly and walked back apprehensively.

"I'm sorry, Dad," I said.

"No," said Hank shakily, kneeling next to his son. "Don't apologize. "How are feeling?"

"I feel a lot better, now," I said, watching the room fill with sparkles. I looked around and smiled. "Wow," I said as I attempted to get to my feet. My father watched my eyes waver then caught me as I collapsed and lost consciousness.

When I awakened I was lying on a hard bed in a white room. I was in a hospital, though I couldn't remember how I'd gotten there. Through the open door I watched nurses and interns hurry back and forth in the hall. The nurses wore habits. A priest, dressed in black stopped momentarily by the door, consulted a little card, realized he was at the wrong room and waved genially at me before moving on. When I tried to smile, I discovered my face was a mass of bandages.

Next to me was another bed where a little boy was sitting up, coughing phlegm into a paper cup. He looked over at me. "Hi," he said and coughed up more phlegm. "I've got tuberculosis," he bragged.

Down the hallway in a waiting lounge a doctor briefed Hank and Joanie. "Okay, here's what we've got," he began "I think Benny's going to be okay. He has sustained some minor trauma to his brain. Apparently his nose cartilage was shoved back just far enough to bruise the surface. Of course we can't tell for sure, but I don't think he'll retain any permanent damage from this." Joanie felt faint and had to sit down. She broke into a sweat and lit a cigarette.

"You say he's not a bleeder. That's good. If he were, we'd be in a lot of trouble, because I don't know why he's still bleeding. We've got it stopped for now. I have him packed, hoping the pressure will do the trick, but wherever it's coming from, and it must be way up inside, we'll eventually find it and take care of it."

They eventually did. But, it took two weeks and numerous procedures before the doctors were finally able to locate the right blood vessels and cauterize them.

The boy in the next bed didn't really have tuberculosis, but he liked saying he did. For two weeks he and I talked and read to each other and held wheelchair races in the hall. But, the races were called to a halt one day when the little boy rammed into a nun and knocked her to the floor.

Had she been a swearing woman she would have cussed up a storm. But instead she said very little as she struggled to her feet and examined her bleeding shins . . .

. . . "Hey!" said Benny "ever heard Mendelssohn played backwards?"

"Who's Mendelssohn?" my mother asked.

"A composer. You know he died when he was only 38? Oh well, it's not important. Come on, you gotta hear this." I started for the stairs. "Mozart died when he was 38 too, I think."

"I can hear it just fine down here, dear. You go ahead."

"Oh, all right. Hey, Mom. Stan should be here soon."

"I'll send him up to your room when he gets here."

Stan was a classmate of mine. We went to school together. I considered him a friend.

I sprinted up the stairs, taking two, sometimes three at a time.

Where in the world did he get these sudden bursts of energy? Joanie wondered. Ever since his return from the hospital he'd been rushing around like a chicken with his head cut off.

Now, he was listening to classical music. Certainly not to her liking, nor Hank's either. They'd always listened to jazz—Louis Armstrong, Benny Goodman, Chick Webb, anything swing. But classical? Where'd he get that from? And backwards at that!

The doctors had warned her that I might go through some changes because of the injury. The bruising to my hypothalamus was real and bound to have an affect. But eventually, they assured her, I would return to my old self and be as good as new. She wondered. It seemed to her that some of the changes just might become permanent. And, she was right, they would.

At first the changes were subtle—a sudden preference for coffee, a tendency to lapse into long periods of deep thought, or deciding to part my hair on the opposite side—those things she could live with. But, then there were changes that weren't so subtle—explosive bursts of energy, a sudden and profound interest in philosophy, religion and science, and most frustrating, my compulsion to write.

Sometimes, seeing light coming from under my door, she'd find me sitting at my desk, writing. I'd be lost in thought or so involved with the structure of a sentence, I wouldn't hear her enter.

"Benny," she whispered. "It's two o'clock in the morning."

I stopped, my pen still poised. "I know what time it is, Mom."

"Don't you think should get some sleep? Tomorrow's a school day."

"No. I think it's more important that I write when I'm inspired to. Besides, I don't need as much sleep as I used to."

"You should get at least eight hours of sleep every night."

"I only need four, Mom."

"Four!" She whispered louder. "No one can survive on just fours of sleep ."

"Thomas Edison survived on less."

"Well, you're not Thomas Edison."

I sighed and put down my pen. "Would it make you feel better if I went to bed now and you tucked me in?"

"Yes, it would," Joanie admitted.

I left my desk and crawled under the covers.

"Good night, honey," she said after pulling the blankets up to my chin. She gave the room a quick inspection to assure herself that all was well, then kissed me goodnight.

"Goodnight, Mom," I smiled.

Joanie switched off the desk light on her way out. "Oh, Mom?" She poked her head back in. "Would you please ask Dad to buy me a typewriter?"

"A typewriter? But you don't know how to type."

"It can't be all that hard. I thought you could show me where to put my fingers on the keyboard and then I could practice."

Joanie hesitated. No harm in that. "All right," she agreed. "I'll talk to your father about it."

I lay in the dark listening. I waited until I was sure my mother had finally retired to the master bedroom, then slipped back out of bed, flicked the desk light back on and resumed writing.

The next day my father carried a typewriter up to my bedroom while I was still in school. It was a Royal typewriter, and it worked fine. It had just been collecting dust in the storeroom at his office. Joanie followed Hank up to my room, cleared a space on my desk, then dusted the typewriter off. When she was satisfied it was clean, she covered the letters on each key with black electrical tape.

"Why are you doing that?" Hank asked.

"Benny said he wanted to learn to type. This is the way I learned. He shouldn't have to look at the keys to know where his fingers are going."

"Yeah, but—he's only eleven years old."

"He's nearly twelve, but I have a feeling that's not going to make much difference. What's this?" Hank asked, fingering an attachment to the lamp cord.

"I'm not sure, but I think he's working on a way to dim the light on his desk so it won't be so obvious when he stays up late writing."

My mom tried to ignore the changes. At first it was easy. She buried her head in a book and withdrew. If she didn't think about it maybe everything would be all right. After all, it takes time to heal from a physical trauma. She couldn't expect me to act like nothing had happened.

Something had happened—I'd been kicked in the face so hard the doctors were worried I might suddenly lapse into a comma or suffer a cerebral hemorrhage—but, I hadn't. And, I wouldn't. I was okay, now. Well, kind of okay.

I'd been home from the hospital for about a week when I came into the kitchen carrying an open book in one hand and a screw driver in the other. "Have we got a soldering gun?" I asked, looking up for a moment from the page I was studying.

"A soldering gun?" she asked, putting down her book. "Yes. There's one in the toolbox under the workbench in the basement."

"And solder?"

"There should be a little container of it next to the gun."

"Okay. Thanks." I resumed reading and wandered out of the kitchen. "

Joanie retrieved her book, then thought better of it. She heard me descend the basement stairs and then listened as I rummaged through the toolbox. She decided to wait for me at the top of the stairs. Yes, I'd changed alright—always preoccupied now—even as I trudged up the stairs I continued to read my book.

"What do you need a soldering gun for?" she asked as I excused myself and brushed past.

"I'm installing a switch so I can reverse the polarity of the turntable motor on my record player," I answered over my shoulder.

"What for?"

I was now trudging up the steps to my second-floor bedroom. "So I can play my records backwards."

She wasn't sure what I was talking about until about an hour later when she heard the strange sounds coming from my room as words and music played in reverse droned from the speaker. "What next?" she thought, shaking her head in private amusement.

If those had been the only changes, Joanie could have soon learned to ignore or least overlook them. But, something more dramatic was taking place in my brain.

Two days later, I burst into the kitchen clutching a volume of the Encyclopedia Americana.

"Hey mom," I said, laying the book open on the table. "See if I've got this right, will you? Tell me if I make a mistake."

Joanie put down the detective novel she was reading and studied the page laid open before her. I paced the floor and rattled off numbers while she followed along in the book. To her surprise I had memorized the page without error. Though, why anybody would commit to memory the population of every major city in the world was beyond her. When I finished, I stood there, beaming.

"Very good," said Joanie, handing the book back to me. "Was that for school?"

"No," I scoffed. "That was for fun." I stood back and grinned. "You don't get it do you?" I asked. "Here, look at the bottom of this page," I said, turning to a bookmark and lowering the volume back onto the table. "There's a list down at the bottom, see?" My Mom looked. Yes, there was a list of planets, their circumferences, their distances from the sun and their distances from Earth

Beginning with Mercury and working my way to Pluto, I began reciting the information. There was no hesitation, no break between planets. Joanie followed along, studying the fine print at the bottom of the page. Every numerical quote was verbatim. When I finished this time, I put my hands on my hips and smiled.

"Well . . ." Joanie said, not really sure how to respond. "How long did it take you to memorize this?"

"That's just it—I didn't really memorize it. I cheated."

"What do you mean you cheated?"

"If I study something long enough, pretty soon it becomes a picture in my head. Then, all I have to do is look at it—and read what it says. It's cheating—it's not really memorizing."

"How long have you been able to do this?"

"I don't know," I shrugged, picking up the volume and starting from the room. "How do they do it?" I asked suddenly, turning back.

Before Joanie could respond, I hurried on: "how do they know how far away a planet is from the sun—or from each other for that matter? It's not like you can take out a ruler and measure it. How do they know?"

"I'm sure I don't know."

"It's a mystery, isn't it?"

"Yes," Joanie agreed. "It certainly is a mystery," she said, retrieving her detective novel.

"Well," I said, then paused as if lost in thought. I stood for a long time staring at a point in space without finishing my sentence.

Joanie watched me. As time stretched on, she began to worry. What's wrong with him? He never acted like this before. Then suddenly, as if awakening from a sleep, I focused on her and smiled. "I'll look into it," I said with a nod and left the room.

Joanie felt herself break into a sweat. Where was all of this leading to? Were the changes harmless? Or, would everything suddenly backfire and I fall into a stupor and finally a coma? Was I turning into an imbecile or a genius?

Up in my room now, I sat cross-legged on the floor and pulled records from their sleeves and spun them backwards on my turntable.

Across the hall, Jeffery closed the door to his bedroom. He started down the stairs, then for some reason thought better of it and returned to stand in my doorway. I was shuffling through a stack of 45's.

"What are you up to, Benny?" Jeffery asked.

"Oh hi Jeffery. Nothing. Have you seen my Stan Freberg record?"

"Which one?"

"The one that makes fun of that TV show—Dragnet."

"Nope. Haven't seen it. Well, I gotta go. See ya later," he said, disappearing from the doorway. I barely noticed Jeffery's departure.

Jeffery had lived with the Olsteins for little more than a year, and still our conversations hadn't progressed beyond surface amenities. There was no love lost between us—there had never been any to begin with. We barely knew one another yet we each tolerated the other for the sake of the mother we shared.

To Benny, Joanie was just his mother, nothing more. She was the woman who bore him. Beyond that, his feelings for her remained aloof—the relationship never manifesting into a genuine affection. Joanie failed to maintain a physical closeness after infancy, so Benny merely mirrored her temperament. Instead, he found solace in interests outside the home or within the warmth of his mind—a web now stretched taut with pubescent desires and whimsical imaginings.

If Joanie felt strong affections she was incapable of expressing them. Benny never missed the maternal intimacy because he couldn't remember receiving it. Holly was much the same and followed suit. Jeffery, however, imagined he could remember nothing else.

Jeffery's adoration for his mother was profound. Being close to the only woman he'd ever loved had become a requirement for survival. He'd been an incomplete person far too long, been separated from her for an unconscionable stretch of time. Now, he burned with a sense of urgency to make up for that lost time, those lost moments, those lost expressions of affection.

The memory of standing on that snow-packed country road, watching the crest of a hill day after day, waiting for a mother who never arrived had been smothered—neatly tucked away where he wouldn't have face it again. The memory of a snow that fell endlessly from a sky turned ugly with gray, the damnable wind that sent shivers up his back and blew snow down his neck, the cold that turned his toes blue and finally numb, and the tears that fell, but never froze—were put aside, forgotten, vowed never to be drudged up again. He had a new life, now. The past was dead— buried along with the memory of his father.

It was seldom he thought of his father. When he did, the world turned cold, a sinking feeling dragged him down, and he lost his smile. These were memories he had to fight like hell to keep buried. Like that night on the porch when the stars sparkled out of focus through the cold and the angels . . . no! . . . Jeffery shook his head to knock the memories away and almost lost his balance on the stairway when he did.

Behind him the voices from Benny's speaker mumbled and grumbled as he gripped the banister to steady himself. He couldn't allow himself to think about that night. He refused to. He would will himself not to think.

But, it did no good. He felt himself slipping. He squeezed the banister until the joints in his fingers ached, but nothing helped. He was being swallowed—falling helplessly into a whirlpool of memories he couldn't erase . . .

. . . During the brief period he was forced to live in his father's house, Jeffery could only recall rooms lit dimly through lamp shades turned yellow with age, an unfriendly kitchen of cold linoleum and a refrigerator smeared with tractor grease and dried food.

All of the rooms were barren of warmth. His bedroom was spiritless, the walls naked and drab. The back porch was neglected until it was in a state of perilous deterioration. Its rotten wooden steps supported a rickety banister that would soon serve no purpose at all.

Any attempt he made to foster goodwill with his father always seemed to end in disaster. Unwilling to give up, on Christmas Eve, after Jeffery was sure everyone had gone to bed, he prepared to slip a little snowman he'd made into his father's stocking.

"Frosty" was fashioned from three walnuts glued together then painted white, the product of several hours of painstaking work. He used a fountain pen to draw its eyes and mouth.

Earlier in the day, when Jackie wasn't looking, he removed several strands of tinsel from the Christmas tree and later, wrapped them around the snowman's neck for a muffler. He glued a tiny orange button onto the head for a nose then hid the completed project under his bed until he thought everyone was asleep.

He lay awake for several hours listening to the sounds the house made at night. The walls groaned when the wind blew and the screen door banged and banged until Jackie or his father grew tired of the noise and secured the latch. Some of the floorboards complained when he walked on them—at night, they creaked by themselves.

From the kitchen, cupboard doors opened and closed, and pots and pans chimed as they were stored away. Water rushed into the sink, the faucets squeaked and plates scraped over one another as Jackie finished up the dishes.

He heard the hum of the furnace forcing hot air through the floor vents. Eventually, the house quieted. A couple of times he thought he heard his father's muffled voice coming from the kitchen, but that was a long time ago. Now, the house was still.

Jeffery slipped out of bed, tied his bathrobe secure and pulled on a pair of socks. The floors were cold. He quietly opened the door to his bedroom and held his breath as he closed it behind him.

The Christmas tree lights had been left on. There were stockings hanging from a shelf that served as a mantle. It was his first Christmas away from his mother. The ghosts of carolers sang from the past and the memory of his mother's warmth swelled in the agony of her absence. He choked to suppress a sob, telling himself he had to be brave.

He padded to the hearthless mantle to inspect the stockings—all three were still empty and hanging limp. He held the snowman up and admired his handiwork. The reflection of the tree lights turned the figurine tangerine then blue. The tinsel muffler sparkled and moved with the warm air rising from the furnace vent.

Jeffery danced his snowman through the air to the toes of the hanging stockings. They were tacked up high and would require a mighty reach even on his tiptoes, but he thought he could manage it.

"Frosty the snowman," he whispered softly as images of pages from a storybook flipped behind his eyes. His mother used to hold him on her lap while she read the words and turned the pages. Frosty carried a broomstick in his hand and ran through the village square, the children ran after him, laughing.

Jeffery surrendered. He stood poised, frozen in time, holding the snowman aloft while the loss of his mother wrapped around him like a cold towel.

Something alive, with arms like an octopus reached out from the bowels of his gut and he thought he'd scream as the monster tried desperately to get out. If only he could purge himself of this demon of loneliness.

He doubled over in pain as he imagined the monster crawling up his food pipe and lodging in his throat. The tentacles unwound, stretching, smacking the sides of his throat as they worked their way out, gagging him, depressing his tongue, bubbling over his lips. He thought he was going to be sick as the hideous snake-like appendages waved before his eyes.

In time, still digesting his loss after swallowing his tears, he sighed and gathered his wits about him. He wiped his face with his sleeve and gazed at the little snowman. Maybe his father would look more kindly on him Christmas morning when he reached into his stocking and found the snowman he'd made.

On the few occasions when their eyes met, his father's eyes were like a doll's, lifeless and blank, looking but not seeing. Maybe this offering would change all that. His father would be so tickled he'd pick Jeffery up and hug him and tell him what a good boy he was. "I love you, little guy," Jeffery imagined him saying. "I'm so damned proud of you, I could bust." Yep, that's just how it was going to happen. Jeffery was sure of it.

He thought he heard whispering. No, not whispering—but, something. Still holding his snowman, Jeffery left the mantle and drifted towards the source of the sound. The kitchen door was closed tight. The lights were off, but he could see something glowing under the door.

Maybe somebody was still up. He got on his knees to look, but all he could see was old linoleum smeared with crumbs and dust. He was sure he'd heard something. Maybe it was the wind—but, that didn't explain the strange light that flickered now and then. Could it be . . . an angel? An angel visiting him on Christmas Eve?

Sure, that must be it! Maybe the angel was making a pie for Christmas morning, just like mommy used to bake for the holidays—all pumpkiny with a thin flaky crust. Sure, that made sense, what else could it be? Maybe there was more than one angel. Maybe there was a whole host of them in there, all whispering so no one would hear them.

He could just imagine Christmas morning. When they walked into the kitchen, Jackie and his dad would stop suddenly in surprise. Resting on the table with lit candles at either end would sit the pie—steam still rising from it.

"Jackie, I didn't know you baked a pie for Christmas."

"I didn't. Where did that come from?"

"An angel made it in the middle of the night while you were sleeping," Jeffery would tell them.

"Don't be silly," Jackie would respond.

"It's true—I saw them. There was a whole bunch of 'em."

Jeffery smiled at the thought. Now, he'd have to find out for sure. The only way he'd know for sure was to see them with his own eyes. And, the only way he'd be able to see them without disturbing them was through the kitchen windows on the back porch.

It was probably pretty cold outside, but it wasn't snowing anymore, so he wouldn't need his hat. He dropped the snowman into the pocket of his bathrobe and left by the front door.

Outside, a steady wind blew from the north. Gentle though it was, the mercury dipped into the twenties. Still in his stocking feet, Jeffery slipped on a pair of rubber boots that were on the front porch and shuddered before plunging his hands into his pockets.

Except for a front of menacing clouds that lurked on the horizon, the sky was clear and the stars twinkled out of focus through a wintry chill. Moonlight fell on the snow and made it easy to see.

When he reached the back of the house, the stairs leading to the kitchen door looked anything but inviting. Beneath a dusting of snow, swollen icecaps clutched the surface of each wooden step. Above, the strange light moved behind the glass. Something eerie was happening in there—something wonderful and exciting. Peaking inside would be like spying on God when His back was turned.

Cautiously, using one wobbly banister for support, Jeffery made his way up the staircase. As he drew nearer, the strange light grew brighter and his heart quickened. With only a couple of steps to go, he slipped and landed hard on his knee. A shock wave of pain shot up his leg, but he stopped himself from crying out so he wouldn't frighten the angels away.

Finally, he reached the porch—sliding a little on the ice and snow, but managing to keep his balance. The worst was behind him. When he inspected his pajama bottoms he saw something dark bleeding through the fabric, but there was no time to worry about that now.

The banister, what little help it offered, had been cold—he may as well have been holding onto an icicle for support. He cupped his hands and blew into them for warmth while scolding himself for forgetting his mittens. Oh well, he'd soon be back inside. Anyway, a little cold never killed anyone.

He moved gingerly over the crusty mantle. When he was close enough he peered expectantly through the glass, his nose touching the transparent cold.

He scanned the length of the kitchen. Nothing by the stove— maybe they'd left. Maybe the angels had finished their work and were already on their way back to heaven. Jeffery turned away from the window and searched the sky.

If they were on their way back, maybe he would catch a glimpse of them as they ascended—their giant white wings flapping in the winter sky—floating up like balloons, getting smaller and smaller. But then, at the last minute, the littlest angel might sense Jeffery's eyes on him and look back. He'd stop his ascent momentarily, floating motionless in space, then smile down from his great height and wave.

Jeffery waved back at the empty sky. "Goodbye," he whispered.

Scraping—like chair legs scooting over a floor made him jump and he turned back. Maybe one of the angels had stayed behind to clean up. Maybe he hadn't looked close enough. He peered through the window again.

The source of the flickering light was a candle. It was sitting on the counter next to the refrigerator making the shadows move like living forms. He looked for the table, where the pie should be sitting. There it was! At least that's what it looked like. Jeffery rubbed his eyes and tried to focus again.

The shifting shapes didn't make much sense at first, but soon he was piecing them together. Yep, that was the table alright, but it was a funny shaped pie. Suddenly it moved . . . or turned. Jeffery held his breath and his heart sped up.

Slowly at first, then with increasing speed, like a ball rolling down a hill, everything began falling into place. The pie opened its eyes. Jeffery stopped breathing. A pie with eyes!

A movement near the end of the table caught his attention. It was his father—he was still up. Jeffery thought his heart was going to pound out of his chest.

Pete reached for a bottle and raised it to his lips. As he drank, it dribbled down his chin and splashed onto his chest. He made a sour face and ran his fingers through his hair. He was naked except for his underwear. He staggered and raised the bottle again, readying for another swig, then stopped, licked his lips and blinked hard. His body swayed as he stared stupidly at the table.

The shape on the table moved, the flickering flame finally lending just enough light for Jeffery to see. The pie with eyes moved and Jeffery jumped in fright. The shape on the table shifted, the pie turned and the eyes blinked as it morphed into a recognizable head. It was Jackie! Her head was on the table. Where was the rest of her? Jeffery squinted through the glaze of the frosted window. Jackie was bent over the table, lying on her stomach—her housedress hiked up to the waist.

Pete took another swig from the bottle then held it over Jackie's recumbent body and grinned as he poured liquor on her back. Jeffery could barely hear them, but he thought he heard her laugh. Pete set the bottle on the floor at his feet and almost fell over before straightening up again.

The warmth from Jeffery's breath fogged the window until it was impossible to see. In an effort to wipe the glass clear, he rubbed it with his fingers. The glass squeaked and he froze. Had they heard him? He listened—nothing—there was no sound at all now.

The sky seemed to grow lighter as the front drew nearer, and moonlight bounced off the clouds. Though Jeffery couldn't tell the difference, the air warmed as the clouds insulated the region and nature prepared to dump a fresh fall of snow.

He peaked back through the window, but the fog had turned to frost, further clouding the glass. Again he tried wiping the window— more squeaking. Finally, the window cleared enough for Jeffery to see again.

Something dark was blocking his vision. He squinted and tried to focus, but even the light from the candle was partially hidden.

"Why, you little goddamned pervert," Pete growled. His drunken face was pressed up against the window, his nose pushed flat like a pig's. The sight caught Jeffery by surprise. He let out a cry, stumbled backwards and fell.

Pete's face was contorted in anger. "You stupid little goddamned pervert," his father yelled through the glass.

Jeffery heard the bolt fall into place as Pete locked the backdoor. "I'll teach you, you little son of a bitch," he scowled. His towering shape disappeared from the door as he stormed from the kitchen.

I'm gonna get it now, Jeffery thought. I'm really gonna get it now. Not bothering to prepare for the leap, he threw himself from the back porch and landed in the frozen snow, rolling until his body slammed up against a snow bank.

Without stopping to consider if he'd injured himself, he was on his feet and running. His rubber boots sank deep into the snow as he made his way around the side of the house and his knee ached where he'd fallen earlier. Finally, he turned the last corner and sprinted for the front door. Just as he reached the top step he heard another bolt click. The lock to the front door had fallen into place.

Jeffery rushed to the door and tried the doorknob, it turned and his heart lightened. He pushed, but the door didn't budge. He was more surprised than shocked, like being slapped in the face. He envisioned spending the rest of his life in the dark and the cold and the snow.

"I'm sorry," he cried. He knew he'd better start apologizing, now—and apologize profusely if his father was ever going to show him any mercy. "I'm sorry! I'm sorry!"

"Fuck you," came his father's voice from the other side of the door.

"I'm sorry," Jeffery repeated, trying the knob again. "Please let me in."

"I'm sorry, I'm sorry, please let me in," his father mimicked, ridiculing him. "Well, fuck you!" he shouted. "You disgust me." His voice had turned bitter.

"I'm sorry," four-year old Jeffrey repeated, his voice breaking, as he tried the knob again.

"Forget it you little pervert. You made your bed, now you can lie in it," he bellowed. "And, it's gonna be an awfully cold bed for you tonight." He laughed. "See if that doesn't teach you a lesson."

The wind was picking up. A few flakes began drifting earthward. Jeffery shivered and his fear suddenly turned to anger. He kicked the door. "Please let me in, I said."

"Oh, you're giving me orders now, are ya? Forget it, stupid. You can stay out there all night and freeze to death for all I care."

"Please. I'm sorry. Please let me in."

"What's going on out here?" It was Jackie's voice.

"Jackie!" Jeffery screamed, pounding on the door. "Jackie, let me in! Please let me in!"

"Jeffery?" Her voice sounded surprised. "Jesus Christ, let him in, Pete. He'll freeze to death out there."

"Well now, isn't that just too goddamned bad," Pete sneered.

Jeffery heard Jackie's footsteps coming closer to the door and his spirits lightened. He even smiled.

"Jeffery?" he heard her say, very near the door now.

"Get away from that door. He's being punished for being a pervert."

"Don't be an asshole, Pete," Jackie snapped.

Jeffery heard the door latch click and breathed a sigh of relief as the door started to open. Suddenly it slammed shut again and the lock clicked home.

"I said he's being punished you fucking bitch."

"Pete, you've had too much to drink. You don't know what you're doing."

"Oh, is that right?"

"Get out of my way, Pete," Jackie insisted. Her voice was very near the door— loud and stern.

"Well now, just who the hell do you think you're talking to, bitch?"

"I'm talking to a drunken fool. Now, get out of my way."

"Forget it, bitch."

"Come on, Pete," she pleaded.

"You can just take your filthy bitch hands off of me."

"You can't leave him out there. He'll freeze!" she screamed.

"I can do anything I goddamned well please." Jeffery heard them scuffling. Then, in the midst of their grunts he heard a loud slap.

"Goddamn it, Pete, you hit me!"

Another smack. "That's right bitch, I hit you. You're a real smart lady. Did you figure all that out by yourself?" Another smack and he heard them retreating. Jackie sounded like she was running for their bedroom. Pete's heavy footsteps followed.

It was suddenly quiet.

Jeffery knocked on the door. No response. He knocked again. Still nothing. He continued knocking until his knuckles hurt and he couldn't knock anymore. His hands ached from the cold and were turning blue. He felt his energy failing. His dad would open the door sooner or later. He just knew he would.

Jeffery sank to the porch floor, and leaned against the door. He stuffed his hands into his pockets searching for warmth, but found the little snowman instead. He'd had it with him all this time.

Well, thought Jeffery, taking it from his pocket and admiring his handiwork, at least I have you to keep me company. Frosty was a comforting friend. Jeffery pressed the figurine to his cheek in an effort to embrace him, then kissed it and smiled contentedly.

As the night wore on, Jeffery settled into an abiding calm. He sat in a quiet daze and watched the snow fall.

A rabbit made its way across the open field. She stopped and back on her haunches, arching her neck as she sniffed the air. When she raised her front legs to her mouth, Jeffery thought she was waving and he waved back. The unexpected movement caught the rabbit by the surprise and she started, but stood her ground. She scrutinized the area until her eyes came to rest on the small human child sitting on the porch of the farmhouse. The child was far enough away to be of no threat, but the rabbit was wary, so kept an eye on Jeffery as she scavenged for food.

To soothe his misery, Jeffery forgot the angels were a product of his wishful thinking. In his body's desperation to calm a mind about to expire, the angels became real—as real as the snow that sometimes swept under the awning and landed gently on his face and melted.

Oh, the angels were real all right. Jeffery had no doubt about that. He closed his eyes and pretended to gaze at the heavens. This made him happy; the anticipation of the angels' certain return consoled him. After a time, he felt his body grow warm. It was no longer as cold as it once was. Obviously this was due to the angels' proximity. They must be very near—floating with godlike precision back to the isolated farmhouse. Jeffery could feel the warm breeze as their flapping wings pushed against the winter air.

His eyes fluttered open. A vibrant shimmering hovered before him—so bright he had to shield his eyes. Gradually, he relaxed as his eyes became used to the brilliant light.

"You're a very special boy, Mister Man," he heard a voice whisper. That's what his mother used to say!

"Mom?" He opened his eyes wide then flinched as the figure before him grew brighter. "Are you an angel?" he asked.

"I'm your own very special angel," the voice whispered in reply.

Finally, when he squinted hard enough, Jeffery could make out the faint outline of his visitor. He was standing with his feet planted firmly apart, towering over Jeffery. His enormous wings arched heavenward, extending far above the visitor's head. The bottom of his wings grazed the porch, brushing the surface of the new fallen snow, scattering it like dust. The wings never stopped moving. They fluttered slowly in graceful cadence—like a butterfly at rest.

"You're here," Jeffery whispered "You're really here."

"Of course I am, my little friend. Where else would I be?"

Jeffery tried to sit up, but his strength had been sapped by his body's attempt to keep warm. So, he sank back and smiled. "I'm glad you came back."

The angel studied Jeffery as the wind pushed his golden hair. It billowed about his head like a halo. "Sing us a song, Jeffery."

"Us?" Jeffery looked closer. Behind the giant angel, hovering overhead, were more angels. "You want me to sing a song?"

"Yes. A song to make the world warm again. We'll melt the snow with your voice."

That sounded like a good idea. Jeffery cleared his throat and began, but his voice was only a whisper and traveled no further than the shroud of falling snow that soaked up his voice like a sponge. Only when the wind shifted, was his voice carried into the open field. In those brief moments, the rabbit stopped munching and watched as the little boy attempted to sing "Frosty the Snowman."

Jackie pressed her fingers against Pete's neck and checked for a pulse. Well at least she hadn't killed him, she thought, returning the oversized alabaster ashtray to her nightstand. Not yet, anyway.

A sharp pain shot through her shoulder and down her arm. That damned ashtray must have weighed a ton. He was gonna have one hell of an awful headache come Christmas morning—that was for damned sure. She'd just tell him he must have fallen and hit his head on a water pipe. Hell, he wouldn't remember anyway; drunken fool that he was—he'd have no reason to doubt her.

She considered lifting him from the floor onto the bed, then changed her mind. She relived the last tortuous hour in an instant—flashing on his fists flying in her face and falling unconscious. She didn't know how long she'd been out, but when she woke up she found herself pinned to the floor, his forearm pressing on her windpipe as he violated her over and over. She kicked his unconscious body hard. "I should kill you for this, you goddamned bastard."

As she got to her feet, Pete groaned. "Fuck you," she said. "I hope you have permanent brain damage, you prick."

She found Jeffery unconscious. His frozen little body was leaning against the front door. She had to move quickly to keep him from falling to the floor when she pulled the door open. She dragged his limp form, kicking the door closed with her foot, and laid him over the furnace vent.

She pulled off his boots then ran to his room and grabbed an extra pair of pajamas. She put his pajamas on a rack in the oven to warm, then went to his side and wrapped him in a blanket. His face was as pale as the moon, porcelain and immobile, as if he were sleeping.

A long time passed before he was conscious again. He was weak, but could feel his feet being rubbed. He let his eyes open slowly, not sure if he was alive or dead, and watched Jackie put his foot between her palms and rub briskly, all the time muttering. At first he thought she was talking to him, but soon realized she was cursing his father.

"Goddamned fool anyway. Lousy son-of-a-bitch. It's not enough you steal the kid away from his mother, but you try and kill him, too. Low-life miserable mother- fucker. It'll be a cold day in hell before . . . Jeffery!" She scooted up and took his head in her hands.

"How're ya doin', hon?" she whispered in his ear.

A strange question to ask; he was unsure how to reply. His first inclination was to say "fine." But, he wasn't fine. He looked into Jackie's eyes and for the first time he saw something approaching concern—almost maternal. How am I doing? "I want my mommy," he sobbed.

Jackie rocked and consoled him as best she could, hating Pete more with each passing moment. None of her sisters would talk to her now, what with the kidnapping and all—and, to make matters worse she found herself living with a complete jerk. Tears streamed down her face as she held Jeffery, trying to comfort him.

The next morning Pete complained his head hurt, but claimed he didn't remember anything—conveniently, Jackie suspected. He said little to his son, except "Merry Christmas," and Jeffery avoided his father's eyes. Jeffery had sustained no permanent physical damage from the cold, which surprised Jackie. She feared frostbite might claim part of an ear or a toe, but it never did and he healed quickly.

But, Jeffery was never the same after that. As the cold swirled around him, he slumped into a dazed otherworld of euphoric contentment as a part of him died.

Like a dying soul floating away from its host, Jeffery sensed a part of himself wasting away—the part that had kept him innocent, the part that wanted to be liked, that enjoyed the company of others, that looked forward to each new day with renewed optimism—the part that allowed him the privilege of seeing the beauty and wonder in all things.

That part died and over the years became callous. Like an untended garden, the damaged virtues were choked out by the weeds of self-doubt and a nagging suspicion he was unworthy of being loved. Tempered caution and distrust grew in its place.

A few months after the trial, while living with his grandparents, the news of his father's death left him more empty than grief stricken. There was some speculation, even by the local authorities, that it hadn't been a hunting accident, after all—but, made to look like one. Jackie was a suspect for some time, but she denied it, and there was no evidence to link her to Pete's death, so the matter was eventually dropped.

Through the years, despite the damage to his spirit, Jeffery remained a gracious man. Girls were drawn to his sensitivity and to the eyes that looked like they were about to cry. He was befriended often by those striving to be near, and though he was courteous, there was always a wall to prevent anyone from getting too close . . .

. . . But, that was all in the past. Things had changed. He had started a new life. He was with his mother now—something he'd yearned for most his life. He was living in a new town, a big city, a new start, reunited. He even had a little sister, Holly, who seemed as untouchable as he, but was slowly warming up to him.

He had a brother—strange character though he was—who at least never competed for his mother's affection. In fact, he didn't seem to need it. Well, why would he? He'd never been separated from her. Jeffery expected Benny's needs to be different.

He liked Benny. Benny kept to himself and never crowded Jeffery or pried. He was always there with a smile, even popping a joke once in a while. And, though they never sat down and really talked, Jeffery felt like he knew Benny anyway.

Yep, It was damned seldom he thought of his father, and now that he lived with the Olsteins he thought of him even less. Instead, he fantasized that perhaps Hank might one day take him aside and ask to adopt him.

He idolized Henry Olstein. Henry was everything he looked for in a father— kind, handsome, strong, intelligent, well liked. Most importantly, he didn't believe in spankings. No matter what mischief his brother Benny might get into, Hank would refuse to raise a hand to him. If there was spanking to be done, it was left up to Joanie, who carried out the sentences with impunity.

Jeffery fantasized about the day when Hank would pop the big question.

"Jeffery," he would say, "can I talk with you for a second. "Listen," he would say, "since you've been living with us, I've come to think of you as . . . well, as one of the family. What I mean to say, Jeffery, is a . . . well, I'd like to make it legal—you know, adopt you." Then he'd pause nervously. "I mean, if it's okay with you. I, ah . . ." he'd search for a word. "Well hell, Jeffery—I love you. That's what I really mean to say. I guess that's it in a nutshell. I love you—and I'd like to be able to call you 'son.' Will you be my son?"

Of course, Jeffery would say yes—but, not right away. He wouldn't want to seem too anxious. Then, if it looked like Hank might be feeling rejected by Jeffery's hesitation, he'd jump right in and say—but not too loud, "You bet . . . Dad!"

Yep, sooner or later it was bound to happen. Jeffery looked to the future with an optimism and trust he hadn't felt since he was a child.

As he descended the stairs he heard the weird noises coming from Benny's phonograph fading behind him. Benny was flipping through a stack of 45's. He was still searching for that Stan Freberg record, 'Dragonet'. . .

. . . "Howdy-doody" came a voice from the doorway.

"Stan!" I said, jumping to my feet.

Stanley Richards swaggered into the room all grins. He was holding several albums in one hand and carrying a satchel in the other. He was tall and slender, but his chest and arms were well-toned from working out with his father's weights, which he did clandestinely when no one was home—which was most of the time. "Guess what I found out," Stanley began.

"What?"

"Did you know that on the exact same day that Galileo died, Isaac Newton was born. Can you believe that? Is that fate or destiny or Karma or what?"

"You gotta be shittin' me."

"No lie," Stanley assured me. "I shit you not. A fact stranger than fiction."

"Wow," I responded, a little dazed. The implication Stan was suggesting was almost too much. "That really boggles the mind, doesn't it?"

"Yeah," Stanley agreed, sitting down. "I nearly flipped my cookies when I found out," he said grinning.

"How did you find out?"

"I was just reading and stumbled across the dates, did a little research and there you are," Stan answered. "Here, look what I brought." He opened the satchel and pulled out several slick covered magazines. "I swiped them from my brother's room."

"Aren't you afraid you'll get in trouble?"

"Nah. He's not supposed to have them anyway. Who's he gonna tell? Besides, I'll put them back before he gets home and he'll never know they were gone. Look," he said, dealing the magazines out like cards. "Gent, Swank, Coronet, Nugget, and Dude."

"Wow," I said, sinking to the floor and staring at the covers. A soft focus of engaging nymphs, their hair slightly mussed, stared back through heavy eyelids and coy smiles. Their partially nude bodies pressed against sheer fabrics of aqua and pink exposing deep cleavages and swollen mounds. "Wow," I said again.

"Truly, magnum opus," said Stanley, staring too."

"Truly," I agreed.

Magnum opus. It's Latin. It means great work. And, these are truly masterpieces of the female form.

"They're girlie magazines. Girls with tits." "Not an easy acquisition," said Stanley.

"You've got to be 21 to buy these treasures," I added.

"To you they're just girlie magazines. To me, they're works of art." He flipped open an issue of Coronet. An attractive young woman was kneeling, facing the viewer. She held a bath towel to her lower half, clutching it to her stomach, failing to hide her navel. The banner arching over the black and white photo read, "Sex Kitten."

"Sex kitten," I whispered.

As we stared in awe at breasts that seemed to defy gravity, Stanley spoke. "I brought the plans for a Jacob's Ladder with me."

"Cool," I replied.

"You'll have to be careful with it after it's built," he said handing it over. "It could easily start a fire."

"Okay."

"If your little sister touches it when it's plugged in, it'll fry her to a crisp."

"I'll be careful."

"I can get you the iron rods, but it'll cost about six dollars."

"Cool."

They stared in silence, then Benny turned the page. They both gasped softly. "It's too bad we can't get rods made out of gold. They'd really conduct electricity."

"No doubt," I agreed. "I wonder why they don't make phonograph needles out of gold."

"Metal's too soft," Stanley informed him, his eyes never wavering from the page. "The point would wear down too fast. Diamonds make more sense. They're harder, lighter and do the job exceptionally well."

"Very true."

"Except they cost an arm and a leg."

"Let's go to the store—and get some Fudgsicles."

We stopped by the kitchen on our way out. Joanie was sitting at the kitchen table. Jeffery was sitting on her lap, again. "We're going to the store. You wannna come, Jeffery?"

"No, thanks," said Jeffery.

"Why don't you go ahead, Jeffery?" his mother urged. "Why don't you go with the boys? I want to do some reading."

Jeffery shrugged. "Okay," he agreed. As he climbed off my mother's lap I noted something odd in my mother's eyes— relief maybe.

Me, Stan and Jeffery walked three abreast up Maple Street. "I'd like to get a copy of that new book, <u>Lolita</u>," Stan announced.

"I saw my dad reading that," I said.

"What's it about?" Jeffery asked.

"It's pornography disguised as literature," Stan said.

"No, it's not," I disagreed. "My dad says it's a work of art.

"Well, forget it then," said Stanley. "I guess I'm not really that interested in reading it, then." I laughed at Stanley's deadpan humor.

Up ahead, two boys, tall and lanky, were leaning against a telephone pole sharing a cigarette. Both were clad in blue jeans that rode too low on the hips and black motorcycle boots. One nudged the other as we approached.

Neither Stanley nor I foresaw an impending conflict, lost as we were in a world of our own, but Jeffery tensed, sensing danger and scooped up a rock, hiding it in his fist.

"I'm telling you, Buddy Holly is the greatest," Stanley insisted.

"The greatest? He's okay, but nothing compared to Elvis."

"Are you out of your mind? Elvis is just a hillbilly hick."

"A hillbilly?" I scoffed. "He's the King of Rock and Roll royalty."

"Rock and Roll royalty?" Stanley giggled. "Don't make me laugh."

"Hey!" said one of the lanky boys, pushing away from the telephone pole. "Where you turd brains think you're going?" He flicked his cigarette into the street with one hand and slipped a toothpick into his mouth with the other.

"Crazy, if we don't change our ways," Stanley quipped. I chuckled at the comeback.

The other boy stepped onto the sidewalk and blocked our way. The three of us stopped in our tracks. "Can't get past without paying a toll," he said.

"A toll?" I asked, realizing suddenly that this was not going to be a friendly encounter.

"Yeah, a toll. You ass-wipes know what a toll is, don't you?"

"What if we don't have any money?" asked Stanley.

"Hey, Buzz, what about that? What if they don't have any money?" he asked.

"Well, " said Buzz, sauntering to his companion's side. "If you don't come up with the money, then we'll just have to kick your ass up between your shoulder blades. Do I make myself clear?"

"Perfectly," Stanley replied, reaching into his pocket. Giving up his allowance was easier than getting beat up. He sure had been looking forward to that Fudgsicle, though.

"How much?" I asked.

"Two bucks," one answered.

"Two bucks," I repeated, reaching into my pocket.

"A piece," Buzz added.

"A piece? We don't have that kind of money," Stanley protested.

"Well, you better find it and quick or this is not going to be your lucky day," said one, cracking his knuckles.

"Now, hold on there," said Jeffery, stepping forward. "I might just have enough for all of us," he said slipping his fist into his pocket.

Buzz and his partner grinned at each other. Stan and I said nothing, but I wondered what Jeffery was up to.

"You got change for ten?" Jeffery asked, moving closer to the two strangers.

"A ten?" one asked, surprised. He chuckled. "You won't need any change."

"I won't?" Jeffery asked innocently.

"Nope, 'cuz the price just went up. A ten will do just fine."

"Okay," Jeffery agreed, acting a little stupid and compliant. He walked up to the taller of the two, the one who seemed to be the spokesman. "You must be the leader," he said to Buzz, grinning.

"Yep. That's me. I'm the guy in charge."

Jeffery stood his ground, still grinning.

"Ten bucks, buddy," said Buzz, his voice dropping an octave.

"That's a lot of money, ten bucks," said Jeffery. "I'm surprised you can count that high," said Jeffery, losing his smile.

"Are you looking for a knuckle sandwich?" Asked the other, smacking his fist into his hand.

"A knuckle sandwich?" Jeffery asked, taking his fist out of his pocket. "I don't think so."

"No," piped Stanley. "We've already had lunch."

Jeffery turned to Stan and smiled. "Hey, Stanley," he said. "Better watch your mouth or I'll tell them you're carrying a twenty."

"A twenty!" said Buzz, pushing past Jeffery. "Holding out on me, huh?"

"I don't have a twenty, for Christ's sake," Stanley denied, backing away. "What are you talking about, Jeffery? Are you trying to get me killed?"

That's when things happened fast. For me, it happened so fast it didn't seem real. Jeffery gave no indication of his intentions. There was no bluffing, no talk—no bullshit.

As Buzz brushed past, pushing Jeffery aside with his shoulder, Jeffery fell in behind him, and with one swift movement, jerked Buzz's pants down to his ankles.

"Hey!" Buzz shouted in surprise. Instincts took over and he unwisely bent over to pull up his pants. When he did, a swift and powerful kick to the back of the knees caused him to buckle and tumble to the ground.

Buzz was attempting to regain his balance when his head snapped back and he blacked out. The toe of Jeffery's shoe had connected with his chin. I watched Jeffery's body leave the ground as he delivered the blow. It was as if he was kicking a field goal. Buzz was unconscious before his head hit the sidewalk with a dull thud.

Before his partner could react, Jeffery turned on him, unbuckling his belt and pulling it free as he advanced.

"Hey," the sidekick said, backing off. "I don't want to fight."

"You should have thought of that sooner," said Jeffery swinging the belt in a tight circle overhead. The brass buckle whistled through the air, sparkling in the sun.

It didn't take long for the sidekick to decide on a course of action. With barely a glance at his fallen partner, he turned and ran.

"Holy shit," Stanley whispered. "If I hadn't seen it with my own eyes, I wouldn't believe it."

Jeffery rethreaded his belt and tossed the rock aside. "Guess I didn't need that after all," he muttered. He started down the sidewalk, lost in thought, then stopped to look back. Stanley and I were staring at Buzz, sprawled on the sidewalk, not moving.

"Do you think he's dead?" Stanley whispered.

"Nah," I said. "He's still breathing."

"Hey, you guys coming or not?"

We caught up to Jeffery, both talking at the same time. "My heart's beating a mile a minute," Stanley confessed.

"I think mine's beating faster," I said.

"Not possible," Stanley disagreed. "I almost had a heart attack. Hey," he said, turning to Jeffery, "you gotta let us buy you a Fudgsicle."

"No, that's okay."

"You saved our lives," Stanley explained. "It's the least we can do."

"We insist." I said.

"Yeah. We insist."

We walked on in good spirits. As we neared the corner, music poured from an open doorway several houses up.

"'Love Is Strange'," said Stanley as the guitar whined.

"Mickey and Sylvia," I said, identifying the artists as the music grew louder.

"Groove Records," Stanley added.

"Blue label, white print," I said as a girl came out of the house and onto the porch. She was wearing a cotton halter-top and tight shorts. Even from this distance I could see her white tummy and the folds on her rump pooching out of her shorts. She was singing along with the song and didn't miss a beat, even when her eyes locked on mine and she smiled.

"Who's that?" Stanley asked.

"Donna," I said, smiling back.

"That's Donna?" Stanley watched her glide to the railing and pick up a sleeping kitten. She held it close to her face and caressed the soft fur while swaying to the music. "I wish they'd allow girls to dress like that in school."

"Not in our lifetime."

"I heard she flunked twice," said Stanley, dropping his voice to a whisper.

"So?"

"Nothing, except that makes her a couple of years older than us, even if she is in the same grade." Stanley felt himself being tugged in her direction. It was almost impossible for him to take his eyes off of her.

She had eyes like magnets, large and inviting. Eyes that suggested she knew something we didn't, but might be willing to reveal whatever it was—if only we had the guts to ask. "She's definitely got the hots for you," said Stan, thinking aloud.

"No way."

"She's smiling at you," Jeffery pointed out. "I don't know if she's got the 'hots' for you, but I don't see her looking at anybody else. She's definitely attracted."

"Yeah, right," I said, wanting to believe it, but not wanting to be disappointed if they were wrong. "I'm too young for her." That's when it dawned on me that Nellie was a couple of years older. And, that never seemed to make any difference.

"We'll see," said Stanley, wishing Donna would look at him the way she gazed at Benny. "Time will tell."

Stanley was right. Time would tell. In the next four years Donna would blossom. By the time I entered high school, Donna would already be a woman—with considerable experience. A condition that would become obvious to me as we shared the same table in Biology class.

But, for now, I lay on my bed with my eyes closed, wishing my headache would go away. I knew in time it would pass. Eventually, the aspirin would take effect, and I'd be able to go back to my writing.

I forced myself off the bed and sat in a chair by the window. The sun was going down. No glorious streaks of color, just a sky that got gradually grayer. My head throbbed. The aroma of pot roast and potatoes, the evening supper, snaked under my door as rain pattered against the window and I wished it were summer again.

The Renton suburbs were barely visible through the industrial haze. A gray overcast loomed above a colorless flight line. At the northern end of the airport, along the pristine shores of Lake Washington, I sat alone in the cockpit of a Cessna 182, bobbing in the water. The plane had been fitted with pontoons and held a full tank of gas.

I fiddled with the antenna of a portable AM radio until a signal finally found a loophole in the atmosphere. "K-J-R, Se-at-tle, Channel 95," a choir of women chanted just before the DJ broke in. "All right all you Beatle maniacs out there—I have what you've all been waiting for, and I've got it right here in my grubby little mitts. It's called 'Something New,' and I'm going to play the entire album for you in exactly 60 seconds—right after these messages." A commercial blared and I turned down the volume.

In the distance, across the gray water at the southern end of Mercer Island, cranes were lowering hydroplanes into the lake. Miss Budweiser was the definite favorite and small crowds had gathered to watch her launch. Behind me, the Renton airport hummed with activity while I waited patiently for my flight instructor.

Planes took off at regular intervals along the single asphalt runway known as 15-33. They buzzed overhead via the east or west channel routes, their engines crackling with interference over the public radio,

A DJ's voice sputtered and I shifted the radio's locational. Suddenly the Beatles broke through clearly for a moment, and I smiled as their music filled the cockpit,

A Piper roared overhead and their voices faded out.

The airport was used predominately by single-engine piston aircraft, and was the training ground for wanna-be pilots. For me, this would be my tenth flying lesson—thanks to my grandfather Leo . . .

. . . Late in the spring of 1964 Grandpa Leo, Grandma Sara and Aunt Fran made their way by train out west and arrived at the Olstein homestead during the first week of June. It was a Friday. Just one week before the last day of my junior year of high school.

On that day, I took a detour after class to walk Donna Piper home. I'd stayed longer than I intended, and later, when I came upon my grandfather, it was with an urgent need to empty my bladder.

I burst into the bathroom without knocking, thinking it was empty. Leo was sitting on the toilet with the seat down polishing his shoes. The rhythm of the swiping brush paused for a beat while he looked up at my anxious face.

"Hello, Benny," he said, picking up the beat again.

"Hi, Grandpa," I said, out of breath. "I gotta go," I pleaded, running in place to keep my muscles from relaxing. Leo moved from the toilet seat to the edge of the tub and continued to brush while I climbed onto the commode. "Sorry," I apologized with a sigh.

"You shouldn't wait until the last minute like that," Leo advised. "It's not good for you."

I nodded, not necessarily agreeing, but too satisfied with the pleasure of releasing my urine to argue.

"Why don't you stand to pee?" Leo asked.

"Mom doesn't allow it. She says we piss all over the toilet seat and the floor and she doesn't like having to clean it up."

"It's not natural. A man can't empty his bladder all the way if he sits like a girl. He's supposed to stand. It's the way God made us."

"Do you believe in ghosts," I asked, changing the subject abruptly.

Leo skipped a beat, but quickly picked up the rhythm again and eyed me. "I'm talking about bladders and you're talking about ghosts."

"Well, do you?"

"I can tell already this is going to be an interesting summer." He considered me and put down his brush. "Ghosts."

"Yeah. Do you believe in 'em?"

"No," said Leo, picking up the other shoe and examining it.

"Not even a little bit?"

"Nor even an iota. Not even a smidgen."

"Why not?'"

"Why should I? Who got time to believe in such nonsense?"

"How do you know it's nonsense?"

"How do you know it's not?"

"Don't you think there might be some truth to the idea?"

"Not even a grain of truth."

'Houdini believed in ghosts.'

"Houdini was a fool."

"He was a great magician."

"He took foolish chances—that makes him a fool."

"Thomas Edison believed in ghosts."

"Edison was a genius of invention—that doesn't make him a genius of anything else."

"Well, what do you believe in then?"

"I believe in you, Benny." He patted my knee and smiled. "Don't worry about what other people believe in—most of them are idiots anyway. Be original. Listen to your soul. So, why are you so late coming home from school?"

The question took me by surprise and I hesitated, searching for an answer.

"It must be a girl," said Leo, nodding knowingly. I felt my ears turn red. "Believe it or not, I too, was young once."

"You're right. It is hard to believe," Benny teased.

"Oh, I know—your generation thinks they invented romance. Believe me, romance was around long before your day, or mine either."

Romance? I wouldn't have called it that. My relationship with Donna had developed into something, but you couldn't call it romantic—alluring, provocative, and bordering on steamy—but, not romantic. Though we'd known each other throughout most of our high school years, and found each other attractive, neither of us had worked up the nerve to actually approach the other. At least, not until today . . .

. . . It was during the last class of the day when it began. I had located my lab table near the rear of the science room and shoved my notebook under the chair. I wouldn't be needing it—I never took notes. Never found them necessary. The only reason I carried a notebook around was because everyone else did. I didn't want to be conspicuous.

I sat quietly and watched the other students file into the room. Stan soon entered and took his seat at a neighboring table. The teacher, Mr. Spencer, was standing behind a long lab bench thumbing through a science magazine, waiting for the bell to ring. The only time he glanced up for any length of time was when Donna entered the room.

"Oh my God, I think I'm in love," Stan whispered.

Donna was definitely a head-turner. She was taller than most students, which was intimidating to some. I didn't know anyone who wasn't a little afraid of her for one reason or another—the girls because she looked and acted more like a woman than a teenager, the boys because her height and build seemed beyond their reach, beyond their experience. Rather than facing the risk of appearing foolish, it was easier for the boys to pretend to ignore her while secretly wishing to be picked up and crushed against her ample bosom.

She wasn't afraid to flirt—in fact, she seemed to thrive on it. She made eye contact with me as she made her way to the back of the classroom. I watched the other boys sneak a peak at her as she sidestepped between rows.

Stan kept his voice low, talking out of the side of his mouth, giving me a play by play of Donna's winding course through the classroom.

"'Hello, Bobby. Excuse me. Whoops, I'm sorry did I just smack you in the head with my tit? Are you okay?' 'Duh, yeah, I think so. Would you mind doing it again? 'Excuse me. Let me just slide by here.' Look at Roger, his eyes are about to fall out of his head. 'Oh, hi Roger. Mind if I scrape my ass against your desk?' 'Duh, no. Not at all. How about if I lay my head on my desk and you sit on it for awhile—would that be okay?' 'Not right now, hon, gotta run, but here, let my just poke you in the eye with one of my nipples.' 'Ow!' 'Oh, I'm sorry did that hurt?' 'I think I'm blind in that eye now, but that's okay, I don't mind.'

I suppressed the urge to laugh too loud. "Wanna trade seats?" Stan asked. "Hey, I'll even give up my seat and just sit on the floor under your table. How about that? I'll sit at Donna's feet and supplicate myself. I'll pray fervently and give homage to her mound of Venus. I'll even ..."

"Hello, handsome," said Donna, sliding into the chair next to me. I was never sure how to respond, so I just smiled and blushed. "Did you miss me?" Again I didn't reply. "Did ya, hmm?"

"Yeah," I finally answered and laughed nervously.

"Good boy," she said, taking off her sweater and hanging it over the back of her chair. "That was the right answer." I watched her caress the mohair, then let my eyes drift to her bosom.

The bell rang and the class made a halfhearted attempt to settle down. Late arrivals received annoying glares from Mr. Spencer who took attendance by checking off names in a little green book.

A dampened hubbub tittered through the science lab. It was never completely quiet in class; students removed notes from pee-chees, snapped three-ring binders, and scooted chairs about. Currently, yearbooks were being passed, contributing to the chatter.

Through the windows I gazed at the marble blue sky. It was spring—and sunny for a change. Only two months to go. Come August I'd be kicking down a gravel road near Irma and Harold's once more—maybe stooping to retrieve a willow branch and whip it through the sweet country air just to hear it whistle. A calmness swept over me and I sighed with pleasure.

Then there was Nellie—a woman now, but just as wild and unpredictable as ever—which was apparent from her letters. We'd been corresponding regularly for several years now. She'd be graduating next week. I wondered what she planned to do with the rest of her life.

"Okay class," Mr. Spencer began, turning his back on the students and chalking words on the blackboard. "The word for the day is deoxyribonucleic acid. Write it down—you'll be expected to know what it is and be required to spell it correctly next week. Wednesday. That's the day of your final exam."

"Shit," Donna muttered, squinting at the board and copying the words. "I'll never remember that." She glanced at Benny who slouched in his chair with folded arms. "Aren't you going to write it down?" she whispered.

I shook my head and smiled.

"Must be nice to be a genius," she muttered. When she finished copying the words she sat back. Donna hated school. What the hell did she need to know about some stupid old acid for anyway? Who gave a shit? She'd much rather be strolling home, or listening to music in her living room. If she hurried, she'd have about two hours before her mother got off work. That meant she could turn the music up as loud as she pleased—even practice stripping to the beat, or at least masturbate for a while.

Mr. Spencer's voice droned on. It was easy to tell that he was fascinated with the subject, but he was incapable of transposing that same enthusiasm to his students. Donna doodled while he spoke, not bothering to take notes—she wouldn't have studied them anyway—so why bother?

Benny seemed interested, she noted. Between doodling and daydreaming she watched him. Man, was he gonna be handsome when he grew up, or what? "Mmm," she said without meaning to, he looked over and she felt herself blush. Damn! She scolded herself. How did that escape? Benny must think I'm a real ditz.

Eventually, even she forgot about the audible faux pas and found herself wondering if Benny was in the least bit interested in her. Well, there was one way to find out. She decided to write him a note—maybe he'd bite, maybe not. Nothing ventured, nothing gained, she told herself.

I felt a nudge as Donna bumped my leg to get my attention then passed a folded sheet of paper.

"This is a model of what we think the molecule looks like," Mr. Spencer continued, approaching a multi-colored structure that looked like it was put together with Tinker Toys. Mr. Spencer stroked the model as if it were alive. "DNA," he said in awe.

I unfolded the note.

"A molecule of this precise structure was discovered by Crick and Watson and contains information that defines every characteristic about you—from the shape of your nose to the size of your heart—and it resides in the nucleus of every cell of your body."

"Not quite," I said softly.

"What?" Donna asked in a whisper, scooting closer.

"There's no DNA in red blood cells," I said. "They don't have a nucleus, therefore they don't have DNA." I dropped my eyes and read the note.

Hey Lover,

I need a big favor.
I hurt my wrist in gym class today. Clumsy me. Do you think you could find it in your heart to carry my books home from school after class? I could sure use the help.

—Damsel in Distress

I looked up from Donna's note into her pleading eyes. I nodded. "Sure," I whispered. Donna breathed a sigh of relief and sat back, content. The gods must be smiling on me today, she thought.

It wasn't what I had been planning. Before agreeing to Donna's request, I was considering catching a bus and going downtown. My father was trying a case that week and I was looking forward to watching him in action.

On the other hand, walking Donna home after school could prove to be mighty interesting. She was, after all, nothing less than fascinating. I tried to imagine what might happen after I walked her home.

She might suddenly lock the door, tear off her blouse—buttons flying everywhere—then grab my head and smother me with her breasts. That was definitely a possibility, I prayed. If only that should happen, I thought rolling my eyes heavenward, life would be complete and I could die a happy man.

Anyway, however unlikely, I'd already agreed—and there was no backing down now.

Still, as class drew to a close, I couldn't help thinking about that big old courthouse . . .

. . . The King County Courthouse was buried deep in downtown Seattle near skid row. Electric trolleys hummed past while pigeons milled about in front, pecking at the sidewalk. Just around the corner on the lawn next to the building, the bums and the hobos slept undisturbed.

Tucked away in a little corner near the entrance, was a dingy cigar store. On his way to court, Henry Olstein always made it a point to stop and visit with the proprietor, reintroducing Benny each time. "You've met my son, haven't you, Bill? This is Benny. Benny, this is Bill." Every time it was the same thing. After a while it became a kind of joke. At least Hank thought it was funny and laughed aloud, patting my shoulder.

Along with Hank's successful career a newborn appreciation for life became apparent and his sense of humor blossomed. "Well, since we're here, I guess we'd better stock up. What kind of cigars are you smoking now?" he asked, turning to me. "Oh, that's right, you don't smoke yet." He laughed at his own joke. "How ya doin', Bill?"

I stood idly by while my father spoke with the shop owner. They'd been chatting for some minutes when an elderly woman entered the establishment. She used a cane to assist her walking and forced a smile when the proprietor greeted her. She wore a worn dark overcoat of wool that rode just above her ankles. Which wasn't odd in itself—it still rained incessantly, but the days had turned warm. Unless she suffered from critically poor circulation, it would have been stifling under those heavy garments.

"A pack of Kool Filters," she said before Bill could ask. She untied the scarf from around her neck and set her purse on the counter.

"Yes, ma;'am," said Bill, pulling a pack from the shelf behind him. "Kool Filers."

"And don't call me 'ma'am'," she snapped, digging into her purse. "I'm a respectable woman and expect to be treated as one."

Hank coughed and averted his eyes while I suppressed the urge to laugh.

"That'll be thirty-nine cents plus a penny for the Governor. Forty cents," said Bill ringing it up.

The woman finally found what she was looking for and laid a twenty on the counter, but kept her finger pressed hard on it. Bill attempted to pull it free, but she resisted, holding it back. "I don't want a soft pack. I told you I wanted a hard pack."

"I guess I didn't hear you," Bill smiled. "A hard pack," he repeated, returning the soft pack to the shelf. "I have one hard pack left." He placed it on the counter. The woman released her hold on the twenty, and Bill made change. "Nineteen and sixty," he said, counting it out.

"It's dented," she said.

"Pardon me."

"This pack. It's dented. I don't want it. I want a fresh one."

"Oh," said Bill, craning his neck for a closer look. "Only slightly dented," he assured her. "I'm sure the cigarettes are just fine."

"Well, I certainly don't want a dented pack" she said slapping it back down on the counter. "I'll take another."

"Bill sighed. "I'm sorry. There isn't another."

"What?"

"That was the last hard pack of Kools I have."

The woman paused. "The last pack?" Bill nodded. "Well, give me back my money, then."

Bill hesitated, then tightening his lips, opened the till, removed forty cents and laid it on top of the other change.

"What's this?" the woman demanded.

"Your money back," Bill answered.

"I gave you a twenty. Not all this miscellaneous change. I'll take back my twenty," she insisted.

"Very well," said Bill. He returned the money to the cash drawer and handed over a twenty. "There you go," he said. "Sorry."

The woman turned to leave, then stopped. "This isn't the twenty I gave you."

"Pardon me?"

"I said," she said raising her voice. "What's the matter with you? Have you got something wrong with your hearing? I said: this is not the twenty I gave you. I want the twenty dollar bill I gave you in the first place."

"Jesus Christ," said Bill losing his temper and banging a key on the cash register. The drawer popped open and Bill retrieved another twenty and handed it over. "Your original twenty. Now, get the hell out of here, goddamn it."

"Well, you don't have to get snooty about it," said the old woman, turning away and tapping her way to the door. As she was closing the door she stuck her head back in, "And watch your mouth," she said, before pulling the door closed, "you cocksucker."

Time had been good to Henry Olstein. He'd strived for success, worked hard and finally gained a reputation as a good trial attorney. And though many attributed his success to hours of legal study, Benny usually found his father reading books by Checkov and other renowned stage directors instead of statutes and studies in jurisprudence.

He read great speeches aloud, rehearsing them until late into the night. He changed his style of dress and began wearing Homburgs and sleek double-breasted suits styled after Bogey. He let his hair grow longer and grayed the temples to suggest a mien of "wisdom". He wore suspenders like Clarence Darrow and sported a mustache like Ernie Kovac's.

When he spoke he adjusted his inflection for effect and worked hard to overcome faults in articulation. He taught himself to look thoughtful like Tracy and learned to smile like Gable. He became a great actor and a great performer. By the summer of 1964, his appearances in court were an event.

The courtroom was sometimes filled to capacity, but there was always a place saved for me near the front where the old men sat. The old men, the regulars, spent their days watching trials. That's about all they did—and they loved it.

They liked my father, and marveled at his genius when it came to the law. They whispered and laughed amongst themselves, but were always respectful of the courts. and mindful not to draw too much attention to themselves.

The last time I sat in on one of my father's cases, it was downpouring. Rain drummed on the window of the courtroom like a thousand tiny fingers.

"No further questions. Your witness, counselor." The prosecutor smiled confidently at the jurors before taking his seat.

"Mr. Olstein, you may cross-examine," said the judge.

Henry Olstein rose slowly, still consulting his notes. He paused, scribbled something on a yellow legal pad, then smiled at the judge. "Thank you, your Honor." He peeked over the rims of his glasses at the witness, then tossed the notebook onto the table, where it landed with a resounding thwack. A couple of jurors jumped.

I giggled inside. I knew my father didn't need glasses; he used them as a prop— as actors do. The surprise explosion from his legal pad was a way to wake the jurors up, to get their hearts beating. A few shifted in their seats and leaned forward.

"Nurse . . . Capner, is it?"

"Yes, Capner," the witness answered.

"Nurse Capner, you testified you took the blood sample from Mr. Miller. Is that correct?"

"Yes, sir."

"I see." Hank strolled closer to the witness stand while buttoning his suit jacket. "By the way, how long have you been a nurse?"

"A practicing nurse?"

"Yes, practicing. What other kind of nurse is there?"

"Well, I received my degree in nursing in 1939, but I didn't actually begin working as a nurse until 1940."

"Okay. So how long have you been a practicing nurse?"

"Twenty-three years."

"Twenty-three years," Hank repeated, then furrowed his brows and paced for a moment. "And, in those twenty-three years, how many times have you drawn blood for the Seattle Police Department?"

"Oh," said the witness almost laughing. "I couldn't begin to guess."

"Try."

"I have no idea. A lot, though."

"Could we say more than ten?"

"Oh, my goodness, yes."

"Could we say more than hundred?"

"Yes, indeed."

"Could we say . . . a million?"

A few snickers rippled through the courtroom. "Your Honor," the prosecutor whined coming to his feet.

"Miss Capner," the judge cut-in, addressing the witness. "Could you please give us a reasonable estimate of how many times you've drawn blood for the Seattle Police Department?"

""Well," said the witness, rubbing her chin and considering the question. "At least thousands."

"Thank you. Continue Counselor," said the judge nodding to Hank.

"Could we say . . . ten thousand, then?"

"Your Honor," whined the Prosecutor coming to his feet again.

"I would say ten thousand would not be an unreasonable number," the witness said.

"So, it might be little less, or maybe a little bit more?"

"Yes."

The prosecutor sank back into his seat. This guy is just digging himself a hole he'll never get out of, he thought. The more expert the witness, the more credible her testimony. This Olstein character is a real jerk.

"So, you've taken a whole lot of blood, a whole lot of times, and given the analysis of that blood to the police department—which they in turn offered as evidence, in a whole lot cases. Is that correct?"

"Yes," the witness answered, nodding. "That is correct."

"And you drew blood from Mr. Miller," Hank said, gesturing to his client. "Is that correct?"

"Yes."

"You also testified that my client was belligerent."

"Yes, that's what I said."

"In what way was Mr. Miller belligerent?"

"How?"

"Yes. Was he rude?

"Well, I should say so," Nurse Capner said with a huff, shifting in her seat.

"In what way—could you give us an example?"

Nurse Capner looked around the courtroom for help, but found none. "Well, when I told Mr. Miller I was going to 'stick' him, he told me I could stick it someplace else."

A few snickers tittered through the courtroom.

"That's what he said? 'You can stick it someplace else'?"

"No, that's not what he said—not exactly," Nurse Capner said, her eyes growing wide with indignation.

"Well, what exactly did he say, Miss Capner?"

Nurse Capner looked helplessly at the prosecutor then turned back to Henry Olstein. "I said, I'm going to stick you now, Mr. Miller. And, he said, 'I'd like to tell you where else you can stick it, lady.'"

"That's it? That's all he said?"

"Yes. And, I think that was quite enough."

"And so, because of that statement, you testified in court today that my client was 'belligerent'?"

"Yes." Nurse Capner squeezed her purse and pressed it hard into her lap.

Henry Olstein walked back to the defense table and picked up a yellow legal pad and consulted it for a moment, then laid it gently back onto the table. He picked up his glasses. "What do you think Mr. Miller was implying, Miss Capner?"

"I think that should be obvious, even to you," she snapped, giving quick glances at the jury and to the judge.

"Oblige me, Miss Capner. What do you think he meant by that statement?"

"Well, if you must have it spelled out for you. I'm sure he was implying that I could stick it where the sun don't shine." A few in the courtroom chuckled.

"And by 'it'," said Hank, relocating his glasses so they sat on top his head, "he was referring to . . . what? The needle you suppose, or something else?"

"Your Honor," whined the prosecutor coming to his feet. "I must object. Is any of this really relevant?"

"Counselor?" said the judge, nodding to Henry.

"Your Honor, everything that happened to this man while he was in custody matters—I'm fighting for a man's life here."

"It's not a man's life—it's a DWI with negligent driving that resulted in a personal injury," argued the prosecutor.

"It's a man's life!" shouted Counselor Olstein. "This will follow him around for the rest of his life. And, I might add your Honor, it was the state that introduced testimony implying belligerency. I just wanted to clear things up."

"All right, all right," said the judge. Please continue. Objection overruled." Again the prosecutor sank back into his seat, a little miffed.

"I'll withdrawal the question, your Honor." My dad looked over at the prosecutor and smiled. "Miss Capner, would you please explain to the jury exactly what you did the night Mr. Miller was brought into you."

"I took his blood to have it analyzed for alcohol content."

"Yes I understand that, but would you please explain to me exactly how you did that."

"Nurse Capner looked around the courtroom for help, but again didn't find any. "Well, I simply applied a tourniquet, found a vein and inserted a hypodermic needle into his arm and withdrew . . ."

"Did you sterilized his arm before you inserted the needle?"

"Yes, of course."

"With what?"

"Pardon me?" Nurse Capner felt a wave of heat wash over her. She wasn't on trial here! That bastard sitting over there was the one on trial.

"What did you use to sterilize Mr. Millers arm before you inserted the needle"

"Alcohol."

"Alcohol," sighed Counselor Olstein, taking the glasses from atop his head and tossing them onto the defense table where they landed with a bang. "Alcohol."

"Yes."

"And the needle you used to stick him with. Was it sterilized too?"

"Yes, sir."

"With what?"

"Alcohol."

"Alcohol," Hank repeated.

"Your Honor," came the bored voice of the prosecutor as he dragged himself to his feet. "I fail to see what any of this has to do with this case. They're two completely different types of alcohol. Neither will affect the reading of the other. I'm afraid counselor is just trying to confuse the jury by attempting to raise a reasonable doubt where none exists."

"The prosecution will get their chance on recross, Mr. Simmons. Please continue, Mr. Olstein."

"Thank you, your Honor. Now, Miss Capner, you stated that everything was sterilized. Is that true?'

"Yes."

"So, you can safely say the blood you drew from Mr. Miller was clean—without any foreign matter in it."

"Yes. I can say that."

"And after you withdrew the blood, what did you do with it?"

"I put it into a test tube and put a stopper on it."

Henry Olstein walked back to the defense table, retrieved his glasses and picked up a large textbook. "What kind of a stopper did

268

you use on that test tube, Miss Capner?" he asked, snapping through the pages.

"What kind?"

"Yes. What was it made of? You stated this blood which you so carefully withdrew from Mr. Miller's arm was placed into a test tube and a stopper was put on. I would like to know what kind of a stopper you used."

"A plain, black rubber stopper. The kind I always use."

"The kind you always use."

"Yes, of course."

"And, you're sure it's rubber and not vinyl or some kind of plastic?"

"It's rubber. It'd always been rubber. I know rubber when I see it."

"Yes, Miss Capner. You're right. It is rubber."

Nurse Capner sat back smugly and nodded her head, quite proud of herself.

"It's always been rubber," Hank said, and laid the large textbook back onto the table and smiled to himself.

It became one of the most important moments is his career. Even while it was happening, Hank had a tough time believing he'd stumbled upon such a gold mine. He knew what had to be said next, he'd rehearsed it for hours and knew it by heart, and when he finally spoke these words, slowly, he let himself savor every syllable. "Your Honor—I would like, at this time, to move for a mistrial."

"A mistrial?" shouted the prosecutor, jumping to his feet. "On what grounds?"

Hank removed the heavy textbook from the defense table. "Your Honor, may I approach the bench?"

The judge nodded and motioned both counselors forward.

"Your Honor, if I may," Henry Olstein said gesturing to the text in his hands. The judge nodded again and Hank hoisted the book onto the bench. "I'm afraid, your Honor, a grave error has been made." Mr. Simmons, the prosecutor, stepped closer to the bench. "Not an intentional error," Hank continued, "but an error nonetheless. One that will affect many, many cases besides this one. The potential for reversals of decision, and suits for damages for wrongful detention, is staggering."

"Oh, for Christ's sake, Henry," the prosecutor said condescendingly. "Do you think you could be any more melodramatic? What's this all about, anyway?"

Henry Olstein ignored him and continued. "As you can see from these studies, your Honor, results from tests made on blood to determine alcohol content are dependent upon their environment.

"Unreliable readings can result from blood coming into contact with certain elements," he continued. "Most notably, your Honor—rubber. Once blood comes into contact with rubber, a true alcohol reading is impossible. Rubber will elevate the true reading, indicating intoxication when none exists."

"Let me see that," snapped the prosecutor, reaching for the book.

"One moment, Mr. Simmons," said the judge, keeping a firm hand on the binding. "You'll get a chance to see it after I've ruled."

My dad reached into his suit pocket and pulled a test tube out. He placed it on the bench and rolled it to the judge. "I've book marked references throughout the text, your Honor. If you examine them closely you will find that in some States, glass stoppers are now being used, while in others, polyurethane has been ruled an acceptable substitute. Twenty states have already banned the used of rubber stoppers precisely because of these findings."

Henry was still speaking, but the courtroom was a hubbub of chatter, and a few local reporters were already running for the door to find a phone. An old man sitting near Benny found it all to be very funny, and laughed quietly, slapping his knee over and over and shaking his head. "That's quite a Dad you've got there, kid."

Yep, time had been good to Henry Olstein—and, it was going to get even better. The financial benefits to the Olstein family exceeded even Henry's expectations. Soon they were looking for a larger and more impressive house and before long a new car graced their driveway.

Taking a bus downtown to watch my father in action would have been great entertainment, but I had already promised to walk Donna home . . .

. . . Now, as I strolled next to her, loaded down with her books as well as my own, I wondered what the hell I'd gotten myself into. Students who watched us walk down the sidewalk together couldn't help but stare. Donna thrived on the attention. She slowed her gait, exaggerated her swing and pushed out her breasts, as if no one hadn't already noticed.

"Well, here we are," she said, turning down a driveway. A buzz saw screamed through a board of lumber and wood thunked onto more cut wood. Carpenters were working on a neighbor's garage. They hammered and clomped over loose pilings, pausing only to admire Donna as she sauntered down the driveway.

I followed her to the backdoor. The journey had been uneventful. The few fantasies I allowed myself while strolling next to her would have do for now.

"Oh, my God, it's hot," I imagined her saying as she pulled off her sweater and tossed it aside. "I mean, it's really fucking hot," she said, pulling her blouse out of her skirt and unbuttoning it. Now her bra was exposed, her bosom spilling over, bouncing with her. Of course I wouldn't have to watch where I was walking—my sole purpose in life would be the careful and extensive examination of Donna's undulating breasts. "Boy am I tired," I said suddenly in my fantasy. "From carrying my books?" Donna asked, real concern in her voice. "The books? Oh, no—my head is tired from carrying around so much knowledge. I know I'd feel much better I could just it down somewhere." "Here," she said, pulling my head close. "You just rest your head on my titties—that'll made you feel a whole lot better." "Indeed," I agreed. Then, while still walking beside her, I laid my head on her warm breasts and my ear slid into her cleavage.

The buzz saw started up again, ripping through my fantasy.

As Donna climbed the stairs to her backdoor, my eyes wrapped around her calves. They were solid and smooth, muscular without a trace of muscle—as if they'd been molded from clay.

I felt an overwhelming urge to take a bite out of them. Just lean forward, take ahold of her leg like a drumstick, and chomp into her calf. The thought made my mouth water.

"Are you coming in?" Donna was standing at the top of the stairs, holding the door open.

I hesitated, a little apprehensive, before stepping inside. My hands were beginning to sweat. I was just about to wipe them on my pants when Donna slammed the door behind me and sighed.

The kitchen was small, but it was clean—the dishes had been washed and were sitting in a rack on the drain board. A note was taped to the refrigerator. Donna pulled it free, read it, and smiled before tossing it aside.

"You can put the books on the table," she said and walked out of the kitchen. "What time do you have to be home?" she asked from another room.

"Oh—no special time," I answered. I retrieved the note and read it.

Hon—

I'll be a little late. Business dinner. Be home around 7 or 8.

Casserole's in the frig. —Mom

As I entered what appeared to be the living room, music blared from a small portable record player. It was Maurice Williams and Zodiacs turned up way too loud, whining "Stay."

"Dance with me!" Donna beckoned. She was moving with the music, curling her fingers. In my fantasy she'd been wearing a skirt and blouse—but in reality, she was wearing a simple beige dress with tiny blue flowers. The thin cotton dress swayed when she moved.

She made me uncomfortable. And, though it was difficult to pull my eyes away from her's, being nervous allowed me the luxury of avoiding her eyes while ogling the perfect shape of her legs.

Unexpectedly, as if tracking the movement of my eyes, her hands clutched the sides of her dress and began bunching the material, raising her hemline. I looked up as her thighs came into view and saw something almost sinister in her gaze.

I was beginning to feel warm, too warm in fact, but I assured myself Donna's behavior was purely innocent, like a flamingo dancer raising her dress while clicking her heels. Yeah, that's it. Because anything else would be too good to be true and I was determined to keep my expectations in check.

Expectations? There were none. Donna winked and smiled alluringly. Nope, by God, dancing was as innocent an activity as playing a trombone in the school band— it was an art form.

Donna turned her back to me and looked over her shoulder. She knew all the right moves and executed them with practiced seductive charm. Her hips moved slowly from side to side as her hemline climbed higher. I wasn't sure if I was supposed to look at her legs or meet her eyes.

Then, for a brief instant, out of the corner of my eye, I thought I saw her dress climb suddenly—all way to her waist, then fall back again.

Impossible, I told myself. Definitely my imagination. Things like that just didn't happen to me. My mind was playing tricks. Yeah, that's it. My fantasies were running ahead of reality.

Oh sure, I'd seen Nellie do lewd things and I'd even seen her naked—but, that was different—she was family. Donna was a complete stranger. Nellie was . . .

. . . Suddenly, I felt a tug, as though some force east of the mountains had lassoed me and was hauling me back to the country.

A thousand years ago, or maybe it was just last summer, when I was staying with my Aunt Irma and Uncle Harold, again, Nellie cornered me in the barn.

"Close your eyes and give me your hands," she ordered.

I complied. You never knew what Nellie was going to do, but there was never any reason to fear for my safety. Her shenanigans were all in fun, so I closed my eyes and grinned. Now what? I thought.

I felt Nellie take my hands and pull my arms forward. Something warm and firm filled my palms. I opened my eyes. Nellie's T-shirt was raised to her chin and my hands were cupping her breasts.

I met her smiling eyes and heard her say, "I was gonna ask if you wanted to go horseback riding, but it looks like you've got your hands full." She burst out laughing.

I remembered laughing too, but I was pretty sure we weren't laughing about the same thing. Nellie quickly stepped back and pulled her T-shirt down.

She slapped the rear of her horse and chuckled. "Come on," she said, climbing into the saddle. "Hop on up. We'll go back to my place and get you a horse to ride."

We took the trail that ran along next to Hangman's Creek. The rope swing was still there, moving lazily over the water. Nellie nodded to it. "That rope swings whether anybody's on it or not. Jeffery used to say it was a ghost. What do you think?"

I wasn't sure what to think.

"Jeffery says it's the ghost of one of the Murdock boys come back to haunt."

"I think it's just the wind," I assured her. And, as if on cue, a gust blew down from the mountains and whistled through the trough of Hangman's Creek.

"Maybe," Nellie whispered. We rode in silence for a spell, and then, "Jeffery says ghosts can take the shape of the wind and make the leaves rustle just like the real thing."

I didn't want to think about Jeffery. I was trying like hell to forget. I felt my mouth go dry as something lodged in my throat—an ectoplasm of a memory impossible to erase. I felt my stomach twist into knots.

For a moment, I thought I was going to be sick. My hold on Nellie tightened and my breathing grew shallow. "Are you okay?" Nellie asked. She sensed she'd said something to make me uncomfortable, so she changed the subject. "Wanna go fishing?"

"Sure," I answered, relieved.

"No. I mean, really go fishing—like on a fishing trip."

Thank God there was something else to think about now. "I don't know how to really fish."

"I'll teach you," she said guiding her horse off the main road. As we trotted up to her farmhouse three dogs tore out of the barn and raced up the dirt road to greet us. "That's Skippy, Mutt and Stinker," said Nellie pointing each out. The dogs were overjoyed to see Nellie and climbed all over one another to be near. "Skippy, there, pees all over the place when he gets excited."

"I have the same problem, sometimes," I quipped.

The Fairfax farm was large and spread out. The barn was unpainted and the blonde wood was warm from the sun and smelled of linseed oil. Inside, it was spacious and bright; light spilled through the stable windows and a thick carpet of straw covered the floor.

"We'll saddle Paint up for you. He's that pinto over there," she said, holding her horse's reins while I dismounted. "Let's head on up to the house first."

Nellie had to pry Mutt off her leg before she could open the backdoor. He'd been humping her all the way up the back steps.

"Dog's a sex maniac," she said

"We have a lot in common," I retorted, following Nellie into the kitchen.

"Hello Benny." It was his Aunt Doris, she'd just closed the oven door and was holding a muffin tin in her hand. "I was wondering when you were going to get around to visiting us." I hadn't seen her in ages, but she was still just as friendly and beautiful as I remembered. Vanilla, blueberries and butter swam through the air. "Anybody hungry?"

As we sat at the table and spread cold chips of butter over the piping hot muffins, Nellie mentioned the fishing trip and Doris agreed that I should go.

"Benny, you'll love it," said Aunt Doris. "It's way up in the mountains—the Cascades, near Glacier Peak. A little town called Darrington. The Stillaquamish River flows right through the town. The salmon will be running—probably steelhead. We'll be driving up late tomorrow afternoon and get there sometime after dark.

"Your Uncle Blair Kellogg and Aunt Lois moved up there a few years back. The change of scenery does them good, I think—at least it gives Blair a chance to hunt and fish. Anyway they've got plenty of room and you're certainly welcome."

"Steve and Theresa won't be there," Nellie added. "They're both off to college."

"That's right. They live in Pullman now. Anyway, I'll arrange it with your Aunt Irma if you want to go."

The following afternoon, Doris, and Earl, Nellie's dad, pulled into Irma's drive. I climbed into the backseat with Nellie.

"Where's Philip?" I asked as the car reached the end of the drive and pulled onto the highway, heading east.

"Oh, big brother is riding the rodeo circuit this summer."

"Doing what?"

"Rodeo stuff. Steer roping, barrel racing—the usual."

"I didn't know Philip could do that sort of thing," I said, impressed.

"Well, you're never here except for the summers. Philip's a real-life genuine cowboy."

"Listen to her, will ya?" It was her father, Earl. "You'd think Philip was the only cowboy in the family." He turned his head to talk while he drove. "Nellie'd be the last one to tell ya, but she's the real rodeo star of the family."

Nellie quickly looked out her window to avoid my eyes.

"She's too modest to ever mention it," said Doris.

Earl cracked his window and let wind blow into the backseat. "Your cousin there is a champion barrel racer herself."

"The horse does all the work," said Nellie trying to downplay it. Earl slapped the steering wheel and chuckled. Nellie rolled her eyes while the countryside raced past her window.

"She won first place last summer," said Doris, turning sideways in her seat and grinning. "I'll bet you didn't know she'll be competing for the title of Rodeo Queen at the State Fair next month, did you?" asked Doris.

"I probably won't get it. There's a lot of competition."

"Oh, yeah. There'll be a lot of competition, all right," her father agreed. "Ellensburg will be overflowing with rodeo talent. But, you're gonna get it. I've seen you ride. And, there simply ain't no other gal as pretty as you are—I don't see how you can miss."

"Dad's a little prejudiced," said Nelllie

I sat back and grinned.

The air blowing in the open window whipped her hair into an ever-changing sorrel aura. She sure was pretty, I thought. Uncle Earl was right about that.

She felt Benny's eyes on her and shot him a quick glance. He grinned and she felt herself blush. She didn't like the idea of Benny admiring her for her horsemanship and didn't fancy him liking her for any other reason than he had in the past.

Scores of miles sped past their windows as the car climbed higher into the Cascades and the air chilled. Nellie pulled on a sweater and folded her arms.

In some places clouds spilled over the mountains and settled on the roadway, the fog slowing their progress. Other times, the air became so clear Nellie was sure she could see for a hundred miles.

When she looked over at Benny sometime later she was surprised to find he was asleep. She watched him sleep for a long time, wishing he would wake up and smile at her again.

By the time we reached the small town of Darrington, the sun had set and the sky was as black as ink. Earl pulled into a gas station and we all piled out to stretch and use the restrooms.

The gas station had a garage. Inside, a car balanced aloft on a hydraulic jack. A mechanic worked underneath, the scent of gasoline and axle grease weighed heavy.

Just a few steps from the gas station sat a roadside diner. Cursive blue and red neon with the word "Pepsi" dangled from a chain in the window. A few patrons dined at small tables or were hunched over the counter sipping coffee.

"I'm so hungry I could eat a horse. Let's get something to eat," said Earl, and not waiting for a reply, headed for the diner. Doris quickly caught up and took his arm. Nellie and I lagged behind.

A semi swept passed on the highway, its air brakes sighing. The leaves overhead shuddered from the gust and I shivered. "Damn," I complained.

"It's getting cold," Nellie agreed and turned back to the car. "Come on," she said, "there's an extra jacket in the trunk."

"See if this fits," she said, handing me a wool jacket. She closed the trunk and leaned against the car while she watched me fumble with the oversized buttons.

"Here," she said, "let me do that." She helped me with the coat and smoothed my collar. When she was finished, I straightened up and stuck my hands into my pockets. "You look real handsome now," she said.

"Thanks," I laughed nervously. I noticed Earl and Doris had disappeared inside the diner. "You're pretty handsome yourself," I said, making light of her comment.

"Wait a second," she said, stepping forward. "It's got a belt." She reached around my waist and pulled it to the front.

Of course, all she was doing was securing my belt, but the attention was making me really uncomfortable. How close were cousins allowed to get? Weren't there some sort of rules or something?

Now she was smoothing it, making sure it laid flat against the jacket—all the way around, reaching, putting her arms around my waist again—leaving them there this time. "Remember that day in the fishing pond?" she asked, looking up at me. "When we went skinny dipping?"

I nodded. "Sure do. How could I forget?"

"I think about it all the time." I did too. "The water was so warm." It sure was, I thought. "I liked touching you."

"I liked touching you, too," I admitted.

There was a movement of her head and suddenly our lips were touching.

I wasn't sure if I'd kissed her or she'd kissed me. And in the end it didn't really make any difference. Neither of us had expected it to happen, yet both of us were powerless to stop it.

"Jesus Christ, Nellie," I said, pulling away. "We can't be doing this. It's against the law or something."

"No it's not," she smiled. "We can do anything we please. We're not hurting anyone."

"What if someone sees us?" I asked, looking around nervously.

"No one can see us," she said, tugging on his belt and pulling him closer. "Kiss me again, Benny." I hesitated and looked towards the diner. "Just once more and we'll go inside."

This time when our lips met I felt her tongue slip inside. That's when the muscles in the back of his knees relaxed and I closed my eyes. I'd stepped off the edge of a precipice and was falling, sinking, drifting slowly into a netherworld. I could feel myself being loved, and I liked it.

If there's a hell, I thought, I just sure as shit bought myself a first-class ticket. On the other hand, it stands to reason, anything that feels this good, must by nature, be good. I wasn't drifting anymore, I was soaring. I pulled her close and returned the kiss.

"Jesus Christ," I said, a moment later, as Nellie walked me to the diner. "I ain't never been kissed like that before."

Nellie didn't say anything at first. She took ahold of my arm and squeezed. "Me neither," she whispered.

"Shit o'dear."

"You know, Benny, the more you're around your relatives the more you begin to sound like them. For a city slicker, you sure do talk like a farm boy."

"So what do we do now?"

"You don't have to go back to Seattle for another two weeks. We'll just make the most of what little time we have." She felt me shake. "Are you still cold?"

"No, I'm plenty warm, alright." It was fear that made me shake. The fear of what tomorrow would bring. I'd seen it happen before, a thousand times. One day a stolen kiss on the playground, and the next day we'd be strangers—as if the kiss had never happened. But, I had a feeling this was going to be different. At least I hoped to hell it was.

"Grandpa says, don't worry your tomorrows, or they'll get stale like day-old bread and you'll grow old before your time."

I laughed. It was as if she was reading my mind. "Yeah," I replied. That sounds like something Grandpa would say. "He told me never to look too far into the future, or I might get my nose bit off. What do you suppose he meant by all that?"

"He meant, enjoy the moment. Don't worry about tomorrow, it'll get here soon enough."

But, I couldn't help it. After a kiss like that every waking moment anticipated the next. But, she was right—our next moment would come soon enough . . .

. . . Blair Savage Kellogg had bought a piece of land less than a mile from the north fork of the Stillaquamish River. There, the river's swift white-water made a sharp turn near the little town of Darrington and began its rapid journey to the west. To the east, the looming hulk of Glacier Peak cast a long shadow over the green valley and siffilating pines curtsied to the wind that swept down its icy slopes.

We arrived hours after the sun had set. Except for a single lantern hanging on the porch, the world surrounding the cabin was black. How Earl ever saw the marker or found the right turn-off in this dark seemed a miracle to me.

The sky was clear, and with no moon, I couldn't remember a time when I'd seen so many stars. I could hear the roar of the river in the distance and feel the chill rolling off the mountain.

The cabin was warm and large. The hide of a pinto horse covered one entire wall, stretching from floor to ceiling. "He was my favorite," Aunt Lois said, caressing the fur. "It seemed a shame to just bury him or burn him, and I sure as heck wasn't going to eat him or turn him into dog food, so I skinned him. Now I have his hide to keep me company. It's like he never left."

In the kitchen, I paused at iron stove and peered unto an uncovered skillet.

'Try some," said Blair, nodding to the pan. "Are you hungry?

"Not really," I admitted.

"Try some anyway."

To be polite, I did. The red meat simmered in its own juices with very little garnishes or spices. It was sweet and fell apart in my mouth.

"It's been cooking for almost four hours, now. Fresh kill. Got him this morning. Used a bow and arrow"

"Really?" It had never quite sunk in that my Uncle Blair was a full-blooded Palouse Indian. I helped myself to another piece. The meat was savory, unlike any I'd eaten in the past. "What is it?"

"Porcupine," said Blair, beaming. "Not bad, huh? Most people don't know porcupine is good eating. Which is good. That means there's plenty out there. No competition for the game. How's your father?" he asked, changing the subject abruptly.

"Fine," I answered. I enjoyed listening to Blair talk. It was almost like listening to Tonto when he spoke to the Lone Ranger.

"I like your father. Follow me, I'll show you where you will sleep tonight." I fell in behind Blair and followed him to the rear of the cabin. "You and Nellie can share Steve's and Theresa's room. Your Uncle Earl and Doris will be sleeping on the fold-out in the front room."

Blair turned a corner. "It's cooler back here, and it's quieter, too." He stopped before a door and pushed it open. "Steve and Theresa are both fond of rock and roll music—that's another reason their room is in the back."

It was a small room with walls of finished fir. Homemade curtains draped the single window. Beaver hides served as throw rugs beside each twin bed.

"Steve and Theresa share a bedroom?" I asked. I couldn't imagine sharing my bedroom with my sister.

"When we lived on the farm they shared a bedroom, too. They get along good together. I've only had to break up a few fights. Now they're off to college. A bunch of your cousins go to the same college. They all share a house. I don't know if Steve and Theresa still share a bedroom. They might. It wouldn't surprise me." He gave one of his rare smiles, which was barely more than a flicker.

"Where's the bathroom?"

"There's an outhouse out back. Take a lantern with you, the trail is hard to follow in the dark."

"You mean I have to go outside in the dark to go to the bathroom?"

Blair didn't bother to answer, he was already on his way back to the front of the cabin, and I stood staring in disbelief at his retreating figure.

I hated outhouses. The idea of plopping my ass down where something could be living underneath was too frightening to even consider. The only thing I hated as much as outhouses was camping. I was not an outdoors person. I liked hot and cold running water and porcelain commodes that flushed. I liked warm fluffy towels rinsed in fabric softener, down pillows and shower curtains with liners. Thank God they at least had electricity.

While the relatives gathered around the fireplace in the front room to visit and sip refreshments, I slipped out the backdoor and relieved myself next to the cabin. I had no intention of wandering down a path in the dark swinging a lantern, searching for an outhouse.

Moments later, when I stepped back inside, I found Nellie standing in the kitchen waiting for me. Her arms were folded over her chest and it was plain she was in good humor. "And, just where the hell have you been?" she asked facetiously.

"Outside taking a piss," I answered. "You should have been there."

"I'm sorry I missed it."

"It was definitely a sight to behold," I said, closing the door softly behind me.

"I'll bet."

"Maybe next time you can join me."

"I wouldn't miss it for the world. Have you tried this meat?"

"Yeah," I said stepping closer and peering into the skillet. Most of the meat was gone now. "I was surprised how good it was."

Nellie paused to taste the meat. We were each wondering what the other was thinking. Because we were cousins we couldn't openly show affection—when an aunt or uncle passed we gave the impression there was nothing special between us— but, because we were cousins, we were able spend long periods of time together without drawing attention.

"Blair tells me we'll be sharing a bedroom tonight," said Nellie, playfully.

"That's the rumor," I said, reaching into the skillet. I picked up one of the remaining pieces of warm meat with my fingers and brought it to my mouth.

Nellie made sure the coast was clear, then slid her hand into my back pocket and gave my butt a squeeze. "So, when do you want to go to bed?"

I started to swallow down the wrong hold and coughed to keep from choking. Nellie laughed softly as she patted my back, and gave me time to catch my breath.

"That meat's mighty rich," said Blair, walking into the kitchen. "You've got be careful," he said, as Nellie continued to pound my back. He poured a glass of water from a pitcher and handed it to me.

"Thanks," I managed, wiping the tears from the corners off my eyes.

"Well, I'm going to bed," said Nellie. "Goodnight."

"'Night," I replied, wanting to go with her, but knowing I should act disinterested.

"That's well water," said Blair, nodding to the glass I was holding.

"Is that good?"

"I don't know. It just is. In the city you drink water that's been treated. Well water comes straight out of the ground—the way God made it."

"Sounds like it might be better then," I said, taking another swallow.

"Maybe. Nobody knows who God intended it for. It could be contaminated." I stopped sipping and looked up in surprise. "Don't worry, nobody has fallen ill," said Blair, leaving the kitchen, "—yet."

"Yet?" I asked as Blair withdrew. "Are you pulling my leg?" No reply. "He's probably pulling my leg," I said to an empty room. I finished off the water then retrieved the single piece of luggage I'd brought along. I checked to make sure my pajamas were on top so I wouldn't have to fish for them in the dark, then wandered into the front room to visit before retiring.

I put off going to bed until very late, visiting with relatives. I couldn't explain it, but sometimes Nellie frightened me a little. Maybe it was her emotional honesty, I don't know, but whatever it was, I wasn't used to it, and secretly hoped she would be asleep by the time I retired. Finally, figuring enough time had passed, I nervously withdrew and wandered down the hallway.

The bedroom was so dark the neurons firing behind my eyes looked like fireworks for the first few minutes. I had to feel my way to the bed. Nellie was quiet, breathing evenly. She'd probably fallen asleep waiting, I thought as I slipped out of my clothes and into my pajamas. My bed creaked when I sat on the edge to remove my socks and made even more noise when I crawled under the covers.

I let out a sigh, closed my eyes and felt myself smile as my head sank deep into the pillow and my brain grew light. "You're about the most bashful boy I know," Nellie breathed in my ear. Her voice had come suddenly, and startled me. I hadn't even heard her get out of her bed. "You never would have come over to my bed, would you?"

"Probably not," I admitted.

"That okay," she whispered. "I never liked boys who were pushy, I felt her lips brush my cheek and my forehead, then my temples. I felt myself being cherished.

"We probably shouldn't be doing this," I said, trying to put some reason into the night.

"It's like a stampede, Benny," she whispered between kisses. "You know what they do about stampedes?" she asked, crawling into bed with me.

"No," I whispered.

"Nothing. You can't stop a stampede. You've just gotta let it run its course," she said, moving against me.

Nellie loved to kiss and Benny was a good kisser. She could feel his warmth. Flannel pressed against flannel. She pulled Benny close and felt him become aroused. Soon, his movements grew steady and his breathing heavier. She became aware of his scent, savoring it, wanting to drown in it, bathe in it. She filled her lungs, then gasped for more. She tasted his sweat. She wanted to swim in it, have it rain down on her in torrents. She imagined their sweat mixing together, blending, mating, making little sweat babies.

Until then, she couldn't admit it to herself, and had gone to great lengths to avoid thinking about it, but as they approached the summit of their pleasures she pressed her lips to his ear. "I love you," she said. "I've always loved you." When Benny responded, whispering, "I love you, too," she exploded. Not once, but several times, one after the other.

The next morning we were up before anyone else. I opened my eyes to find Nellie standing with her back to the door watching me. "Good morning," she whispered. "I need a hug."

She seemed kind of weepy. I wasn't sure if she was regretting the night before or if she was celebrating it. We hadn't really done anything, he reasoned. At school dances it was called "the grind." It was like standing up and screwing with your clothes on. Dry fucking, that's what my friend Stanley called it. Except last night I'd done it with my cousin, lying down in bed with only our pajamas on. I blushed with guilt.

"Come here," said Nellie, holding out her hand. When I took her in my arms, I felt her hips move against mine, and there we were—doing it again.

Later that morning as we followed the grownups to the river to fish, we made it a point to lag far behind. Whenever Nellie felt brave enough she slid her hand into my back pocket. Whenever I felt brave enough I slid my hand under her coat and caressed her bare back.

"A couple of days ago," said Nellie, "when we were riding over to my place, I said something that upset you. What was it?"

"I don't remember," I lied.

"The swing over Hangman's Creek was moving, remember? I said Jeffery used to say ghosts made it move. Is that it? You aren't by chance afraid of ghosts or something, are you?"

"Ghosts? No. I'm not even sure I believe in ghosts."

"What then?"

"Jeffery."

"What's going on? I've heard a dozen rumors."

"Like what?"

"I heard he ran away once. Heard he quit school and joined the Navy. Heard he must have got a girl pregnant 'cause your mother kicked him out of the house."

"None of that's true." If only it had been that simple.

I hesitated. It wasn't something I wanted to talk about. Hell, it wasn't anything I even wanted to think about; it hurt to remember. But, no matter how hard I tried to forget, I couldn't block the images that burst through the floodgates until I was drowning in them—the curse of a phenomenal memory. And, for a reason even I couldn't name, I felt responsible . . .

. . . It happened on a Saturday. And, like every Saturday around the Olstein household everything was in a state of flux.

Holly was up in her room with one of her girl friends with the door closed. They had the phonograph turned up so loud I could hear the muted voice of Johnny Cymbal singing "Mr. Bass Man" all the way downstairs in the living room. Jeffery had just gotten home and I could hear him in the basement laughing at something one of his friends had said just before trudging up the stairs.

My mom and dad had been in the kitchen for some time, talking in hushed voices. Once in a while my mother's voice grew louder in an exasperated explosion of emotion.

"All right, all right," my father whispered harshly. "I'll do it, goddamn it."

"Hey, Benny!" Jeffery called as he sauntered into the room. He'd let his hair grow long, and though it had lost its tight curl, it still held a natural wave. He combed it back on the sides, making a duck's tail; a platinum pompadour fell over his forehead. Hell, I thought, if I dared come home with a hair-do like Elvis, my mother would have a conniption fit and taken the clippers to me. Being the first-born did indeed have its privileges it seemed.

"Gary here lent me his record," Jeffery grinned, motioning to his friend standing behind him. Gary nodded to me briefly then quickly scoped out the room. "Mule Skinner Blues," said Jeffery, sliding the 45 onto the turntable and flipping on the power.

The Fendermen bellowed into the room the Muleskinner Blues. "Good mornin' Captain!"

I liked the song too, so we stood around nodding our heads and tapping our feet to the music. My father stormed into the room, seemed to change his mind, nodded to me then went back into the kitchen.

What the hell was that all about? I wondered.

My father came into the room again and I caught a glimpse of my mother pacing the floor in the kitchen as the door swung closed.

Hank hesitated, then said something to Jeffery and Jeffery nodded. He told his friend he'd see him later and handed him back his record.

Jeffery glanced in my direction and smiled broadly, as if to say, "this is it,"—as if the most wonderful thing in the world was about to transpire.

The room was quiet now. Even the noise that had been drifting down from Holly's bedroom subsided. Jeffery hurried to Hank who was standing by the window. He looked up expectantly, reminding Benny of a little puppy.

Outside, Seattle's overcast blocked the sun and washed the colors out of anything living. A drab of gray fell over Lake Washington and the arboretum drooped in the damp.

"Listen, Jeffery," Hank began. The words weren't coming easy. I watched him struggle—a rarity for a trial lawyer.

"I've been talking it over with your mother and…" —Jeffery looked in my direction again, caught my eye and smiled reassuringly.

"Well," Henry continued, "I'm afraid it's just not working out. I'm sorry, but you can't stay here anymore."

At first I thought I might have heard wrong—Jeffery did too and we both stood dumb—shaken. But, when we saw the pain in Henry's eyes, we knew the worst had happened.

Jeffery's smile faded.

Sucker punched by the announcement, the words slapped away all expression. A meat grinder slid into his abdomen and began pulverizing his insides. He wanted to fall to the floor and writhe in agony—wanted to suddenly end it all by throwing himself out the window—hoping the end would be painful, that the shards of glass would rip away his misery once and for all.

Maybe he was dreaming. He clenched his fists and dug his nails into his skin. Wake up! He scolded himself. This couldn't really be happening. No, sir. Why, he'd already been through grief, enough for a lifetime. Nothing could be this unfair. Hell, he had plans. Lots of 'em. Like . . . okay, so maybe he didn't have any real plans— but, who makes plans when life is good and all is well?

But, he wasn't dreaming. It was real. What have I done wrong? Why am I being thrown away? Why can't we just start all over and say it was all a terrible big mistake? Why doesn't he take me in his arms and weep for forgiveness?

"I didn't mean what I just said," he'd say, pulling Jeffery close and hugging him. "I don't know what the hell I was thinking of. I must have been out of my fucking mind. Please forgive me, I was such a fool. I love you so much, Jeffery. Say you'll be my son."

"Yes," Benny would chime in. "Yes, that's right. I love you, too, Jeffery. Let's be real brothers."

Jeffery averted his eyes from me. He couldn't look at me now— didn't feel like he ever could again. He didn't know if it was because he was embarrassed or ashamed. If he let me see his eyes then I'd know the pain—and nobody should know this pain.

I couldn't believe what I'd heard.

"Your mother and I think it might be best if you went back to live with your grandfather." It was hard for Henry—one of the hardest things he'd ever had to do. The memory of that moment would haunt him to his dying day. He was accustomed to twisting words to defend his clients. He was a master of images—a poet. But, now he was inflicting pain with words and he felt his own tears burn down his cheeks as Jeffery wept.

"Hey," he thought softly to himself, "it's not me, kid. I had nothing to do with it. It's your mother. She's got some crazy ideas that . . . oh, hell, who knows what's going on in her head? But, that's okay. You go ahead and blame me. Sure, why not? It's not good to think the worst of your mother. It's better you hate me."

Big boys don't cry—it's a rule or an unwritten law or something. They just don't do it. And, if they do, they never cry in front of someone else—no matter what. But, Jeffery had lost control. He didn't know if he was mourning for something he'd lost or if he was grieving for something he could never have. He couldn't stop the tears.

He wretched, unable to suppress the agony as the room tilted and the world came unbalanced. He felt like he was going to throw up. He fell into a chair and held his head, wishing he could vomit up the hurt. Wishing he could puke out all the pain from his miserable childhood in one violent eruption. Spew it onto the floor at his feet so he could stomp on it. Stomp and stamp on it in his shoes like a kid playing in a rain puddle—kicking the offal across the room so it splattered on the walls.

He cried for a long time. At times he wanted to jump up and scream at them. "Why are you doing this to me? What have I done wrong? Is my hair too long? Don't I dress right? I can change that. I can change whatever you want. I can be whoever you want me to be. Just please love me. Don't push me away."

Finally, he stopped feeling and went numb. He had no more energy, no more desire to fight, he just wanted to leave now—just wanted to get away from them all and be alone. It was the way he'd spent most of his childhood anyway. "You'd think I would've gotten used to it by now," he said softly . . .

. . . I was leaning against a tree, standing on a path in the woods that led to the north fork of the Stillaquamish River—a river where Nellie and I and others were going fly-fishing. The river flowed right through the little town of Darrington, where its swift white-water made its rapid journey to the west. The salmon were running—steelhead.

Blair Savage Kellogg and his party had trudged ahead, leaving Nellie and I behind to talk. But, I didn't want to talk anymore, I was tired of remembering.

"We don't have to talk about this anymore, Benny," she said, putting her arms around me. "I'm sorry I asked."

"I hate thinking about that day, Nellie. It eats at me like a cancer. Every time I think of that day I feel so ashamed of my mother and get so mad at my father I can't see straight. Man," I said, shaking my head, "poor Jeff—how can anybody live with that much pain?"

"Where is he now?"

I shrugged. "I don't know for sure. He went back to Spokane, stayed with his grandpa one night and was gone the next morning. A couple of months later my Mom got a letter from him."

We resumed walking, following the hard-packed path that wound its way through the underbrush to the river. "He'd joined the Army. I guess when you're in basic training you're required to send a letter home to let your folks know you're alright, so he sent a letter to mom. You should have seen the look on her face when she opened it."

"Wow, it's hard to picture Jeffery as a soldier."

"Yeah. Anyway, we didn't hear from him for quite a while after that. Then about a month ago Mom got another letter from him. He said he'd been accepted to Ranger school."

"Jeffery's going to be a Green Beret?"

I nodded. There was more to tell, but a lot was better left unsaid. Like how Hank came home and saw the letter and sank into a chair at the kitchen table while reading it and started crying. My mother, who had read the letter and remained dry-eyed, wept when she saw Hank's reaction and went a little crazy for a while afterwards—she even tried to kill herself by swallowing a bunch of pills. But, she was okay now. She'd kind of gotten over it. But, the damage to their marriage was irreparable.

Nellie slipped her arm around me. "I'm going to tell you something, Benny."

"Okay."

"It's going to take your mind off of everything."

"I'm ready."

"I could search all my life and never find a guy as wonderful as you. I don't care if we are cousins, Benny. I love you. And, until you have to go back home, every chance I get, I'm going to lift up my shirt and shove my tits in your face."

I laughed. "Thank you. I look forward to it."

For the next two weeks, whenever we could finagle a moment together, we dry fucked. It didn't make any difference where we were or what position we had to assume— as long as we couldn't be seen, we were doing it.

That was nearly a year ago—last summer . . .

. . . Now, suddenly—standing in Donna's living room with my palms sweating and music blaring from a portable phonograph, I missed Nellie.

Donna was unbuttoning the top three buttons of her blouse. She turned her back to me and moved her hips, swaying like a hula dancer. Maurice Williams and the Zodiacs were still singing.

Donna turned and faced me, pretending to pout. She was lip-synching as she danced. It was like watching a movie and difficult for me to believe I was really there. Her breasts were spilling out of her bra—two perfect swollen mounds that moved with a life all their own.

Still lip-synching, she moved slowly to me.

She was standing so close I could smell her Doublemint breath.

When the needle reached the end of the record it rose, returned to the beginning and started again. Donna took my hands and placed them on her waist as she continued to sway.

"Donna," I began, "most guys would give their left nut to be where I am right now." Donna smiled, flattered. "But, I just can't do it," I said, turning and heading for the kitchen.

"Benny!"

I ignored her call and hurried down the back steps to the driveway. The scream of a buzz saws greeted me from the house next door. Lumber crashed to the floor as Donna called my name again. When I hit the sidewalk I broke into a run.

By the time I reached home I had to pee so bad I didn't think I was going to make it before my bladder burst. It was with a great sigh of relief that I barged into the bathroom only to surprise my grandfather, Leo.

Leo slipped his feet into his shoes and stood to admire the high gloss that now graced the brown leather. "Not bad, huh?"

"Not bad," I agreed, pulling up my pants. "Maybe you'd like to polish mine next," I said, nodding to my sneakers.

"Ha-ha. Very funny man. Maybe you should get yourself a job on the Tonight Show and be a comedian. Oh, by the way," said Leo, opening the bathroom door, "I'll be sleeping in your room tonight."

"You won't be sleeping with grandma?" I asked, following my grandfather to my bedroom.

"She'll be sleeping in the spare room and since you have twin beds, I'll sleep in your room. Your grandmother and I have always had separate beds." I looked at him oddly. "That surprises you?"

"Kind of." I went to my desk and sat. "Not really, I guess. It's impossible to imagine you and grandma having . . ." I searched for the right word.

"Sex?"

"Yeah, well," I answered.

"Your generation—you think you invented sex. Your grandmother and me were having sex long before you were born."

"Oh, yeah. That's how dad was born, huh?"

"You're a real wise guy, today." Leo heaved his suitcase onto the bed and opened it. "I've got something in here for you." He found what he was looking for.

"What's this?"

"I won't be here on your birthday—we're going back before then, so I'm giving you your gift now."

I hesitated.

"Go ahead, open it." While I tore open the envelope, Leo went on. "Besides, classes begin next week. I know, I know, your birthday's not until next month, but this way I'll get to be here when you start."

"Classes?"

Leo examined the many small posters of Picasso prints thumbtacked to Benny's bedroom walls. "I don't mind telling you I had to do some pretty fast talking to get your mother to agree to this."

"To what?" I asked pulling the letter from the envelope and unfolding it. Leo was silent while I read. "Jesus Christ, Grandpa," I finally managed, my voice barely a whisper.

"What ever happened to 'holy cow?' Nobody say that anymore."

"I can't believe this." I sat, stunned. "Flight lessons. I'm going to learn how to fly?"

Leo just shrugged and grinned.

"You're not happy?"

"Happy? I can hardly contain myself."

"Well, you could have fooled me."

I jumped up and grabbed my grandfather and hugged him so hard I thought I heard something crack. "Sorry," I said releasing him.

"That's okay," Leo said, sitting on the bed. "I know it's something you've always wanted."

"This is a milestone in my life. It changes everything."

"A milestone," Leo repeated. "Please don't make me regret this. Just be careful. If anything happens to you I'm as good as dead. I had to talk myself blue in the face to convince your mother it was okay."

"Thanks," I said and hugged my grandfather again. "Don't worry, I'll be careful. And, someday when I'm a chopper pilot flying combat missions for the Army, I'll tell them I owe it all to you."

"Great," said Leo, rolling his eyes. "That's all I need. Your mother hears you say that and I'm dead meat for sure." . . .

. . . The Renton suburbs were a ghostly specter through the industrial haze. Another gray overcast loomed. At the northern end of the airport, along the landscaped shores of Lake Washington, I sat alone in the cockpit of a Cessna 182, bobbing in the water.

Across the water, two hydroplanes had been lowered into the lake. Behind me the Renton airport hummed with activity while I waited.

A plane buzzed overhead, heading east, its engine crackling with interference over my portable radio. The DJ's voice sputtered for a moment, I shifted the radio's antenna and the Beatles broke through again. There was a sudden change in air pressure as "the cabin door swung open and former Army Captain Bruce "Buzz" Calder slid into the cockpit next to me.

"Ready for your flying lesson?" he asked, securing door.

"Ready, Captain" I said, buckling up and adjusting my straps.

"Okay. Start her up. You/re flying solo today."

"Solo?"

"Today, I'm just along for the ride. We'll see how it goes. If I'm satisfied you're ready, I'll sign the papers and schedule you for your flight exam."

"Holy shit."

Buzz smiled at Benny's disbelief. He was tempted to tell Benny what an exceptional student he was, but the fear of building overconfidence prevented him. Overconfidence might lead to mistakes and a mistake in the air could be fatal.

Yet, he couldn't help but marvel at Benny's aptitude. Besides his natural flying ability, he could quote the flight manual like he had a copy of it printed in his head.

The pontoons skipped across the choppy water of Lake Washington, kicking up a fine spray. I eased back on the yoke. Gravity pulled on my stomach and the plane rose smoothly into the air.

"I do believe we're airborne," I announced.

"Good," Buzz shouted over the roar of the engine. "We're cooking now." Flying with Benny was a pleasure. He seemed so at ease at the controls—it was as if he'd been flying all his life. "Take her to 3,000 and head north—45 degrees."

"Aye, aye, Captain," I shouted, tossing a fake salute. The plane banked hard and rose swiftly, leaving Renton behind. Seattle passed below, while above the ceiling prevented even a glimpse of the sun.

"What are you going to do after you graduate, Benny?" Captain Calder asked, lighting a cigarette.

"I don't know. Go to college, I guess. But, that's not for another year or so."

"That's not what I hear."

"What do you hear?" I asked, briefly taking my eyes off the horizon to monitor the gauges.

"I was talking to your father and he told me you already have enough credits to graduate if you want to."

"Yeah, I guess."

"So, why are you stalling?"

"I'm not stalling. I'm just not sure I'm ready to go to college."

"What do you want to be?"

"That's just it. I don't know. I never really gave it any thought. I never had my heart set on being a lawyer or a doctor or a baker or a . . ."

"Candlestick maker?"

"Yeah, right," I laughed.

"What do you want to do?"

"I don't know. Fly. I've always wanted to know how to fly."

"Well, you're already doing that."

"Yeah." I took a deep breath and felt at ease. The horizon was a steady line—yet I could still make out the curvature of the earth. A new sense of well-being swelled within and all seemed right.

Up here, I was in control, the master of the universe. I could finally fly like Superman and soar with the birds. Up here I had a better perspective of the world. Up here it was just me and the sky and an appreciation for the wonder of the planet. Up here I was closer to God—of that I was certain.

"Maybe you should join the Service—fly jets for the Navy—get stationed on an aircraft carrier."

"Maybe. I think I'd really like to join the Army and learn to fly helicopters.

"Choppers are a hell of a lot more dangerous than fighter jets."

I grinned. "Danger is my middle name."

"You've been watching too many James Bond movies. But, I'll tell you, Benny, this old soldier has still got a few connections—friends in high places. I could probably get you into a chopper school if that's what you really want. You just let me know when you're ready and I'll see what I can do."

Later that summer, at the end of August, the U.S. Congress was busy passing the Gulf of Tonkin Resolution,

The world had gone black. His eyes closed at the end of a blink. But he wasn't dead—not yet. He was still conscious. He couldn't see any more, and he couldn't move either, but at least he wasn't in any pain and he had some semblance of awareness. He could be thankful for that; at least he wasn't a vegetable.

Music. Music? Now, why would he hear music being played on a battlefield in Vietnam? It was drawing nearer, growing louder, the guitars rumbling, trying to force their way through a filter of chopping rotor blades.

Chopper blades thrashed, whipping the tall grass into a frenzy, beating against his sides and slapping his face.

Coming in for a landing, now. Lowering. Loud rotor blades whacking. The vibration of a machine. A cold floor.

He'd heard voices, some of them were shouting, but the words made no sense. The chopper hesitated for a moment, switched to a new song then began to lift off. Movement. He was flying again! There was another explosion nearby, but the new music continued . . .

. . . A helicopter pilot might receive a pat on the back or a shoulder squeeze as the soldiers piled into the choppers. Captain Benjamin Olstein was even kissed once on the cheek by a sergeant overwhelmed with gratitude. Manning the controls of a Huey UH-1 was like being everybody's best friend, like being a superhero or the Savior. Here, I didn't have to pretend. Here, I was Superman.

On liftoff it wasn't unusual for me to throw the switch on a tape deck, adding my own unique soundtrack to the war. Coming in for a landing you might hear anything from Gustav Mahler to The Electric Prunes, but on take off, it was always something for the boys. Today, it was "He's So Fine" by the Chiffons.

I had flicked the switch automatically upon takeoff and wasn't even listening to the music. I was concentrating on getting the hell out of there.

The skids of the UH-1 clipped the treetops a little too close and I pulled on the collective for lift.

Below, the American soldiers and their enemy were invisible beneath the thick tapestry of green. Instead, I watched long-armed monkeys leap from branch to tree, frightened by the noise and wind; clouds of exotic birds decked in brilliant plumage burst from the mangroves and scattered.

The helicopter rolled off to the right, picking up speed as it rose. I had just inserted troops into a hot zone and was anxious to get back to base—more than anxious—I was desperate.

My bird had taken several hits and we were losing fuel. Out of the six Hueys assigned to the mission, only four were returning.

B-40 rockets had screamed out of the jungle just as one helicopter was lifting off and the other was still hovering. They both exploded in bright balls of orange along with their crew and passengers.

My chopper should have been nearly empty except for the copilot and gunner, but today I was returning with a full load—most of them wounded or dying, some of them the very soldiers I was supposed to be inserting.

Behind me, one of the gunners, the one they called Crazy Horse, was leaning out of the cargo door, shouting—yelling so loud the tendons stood out on his neck. I couldn't tell if he was screaming in anger, or just happy to still be alive. Maybe it was a little of both. The gunner had removed his helmet so I keyed the intercom to my copilot knowing our conversation wouldn't be overheard.

"He's just scared. Don't let it bother you," I said, banging on the fuel gauge with my fist.

"Why do you think he's hollering like that?" my copilot asked.

"Because he's crazy," I answered.

I looked back. Crazy Horse was hanging onto his M-60 and leaning temerariously over the edge. He hadn't taken the precaution of securing himself with safety belts, and the danger of him falling out was real.

"I can't understand what he's saying," my copilot keyed.

"He's pissed. He's saying 'deja-fuck'. It's a play on déjà vu," I said, ignoring the pools of blood on the floor.

"Meaning?"

"Meaning, we've been here before," I answered. "It was only four months ago that we took the same hill. Now it's back in the hands of the gooks. It's the same thing over and over. We insert—and take it. We leave—they take it back. He's pissed. Come to think of it, I'm a little pissed myself."

Off to the left, at ten o'clock, I watched a puff of smoke break free of the jungle's canopy and spiral skyward. It could have been anything—antiaircraft fire, in which case I was as good as dead—or a Bouncing Betty blowing the lower half of an American soldier apart.

"I just want to go home," keyed the copilot in his Southern drawl. "Go home, hug the kids and kiss the wife. That's all I want. I don't give a shit anymore. I haven't been here 30 days and I'm ready to DEROS out of here."

DEROS. The Date of Expected Return from OverSeas. They used the word "expected" because you could go home sooner if you were wounded bad enough or killed. Too many of my friends from flight school had already gone home early. Unfortunately, most of them in silver coffins.

"I shouldn't even be fucking flying today," my copilot went on. "I can practically guarantee you, right now, that by the time I get back to base I will not be feeling well. No sir, I won't be feeling well at all. In fact, I plan to be very sick for the next couple of weeks. Very

sick. Probably a terrible case of the flu, if you know what I mean."
He looked over nervously and tried to smile.

Behind us wounded soldiers were sprawled on the floor in no
coherent order, tossed in at the last minute just to get them onboard.
One soldier lay unconscious, his platinum hair plastered to his scalp,
part of his head missing, something leaking out. A black soldier was
leaning over him, rubbing his back protectively and talking to him.

Meanwhile, the corpsman was attempting to secure tourniquets
on another soldier who was thrashing and panicking, screaming that
he was afraid he was going to die.

Could be, I thought. Both of his legs were torn off at the knee
and arterial blood was pulsing—squirting all over the cab's interior.

"Hey, what's up with the fuel gauge?" keyed the copilot tapping
the indicator dial.

"I think we've sprung a leak," I replied. "We took a lot of hits
back there."

"Great. That's all I need. But, we're going to make it back to base
though—right?"

"I'm not sure. You better radio in—give them a status report.
We're still about fifteen minutes out."

This wasn't the way it was supposed to be. The day I arrived in-
country, when I peered out the window of the DC-10 as it banked
hard and circled for a landing at Saigon International, my expectations
of war had been painted from television and comic books. The good
guys wore white and always won. The bad guys wore black and always
lost. In fact, the Viet Cong ran around in black pajamas, supporting
this notion.

I watched the needle on my fuel gauge dip lower as the soldier
behind me continued to scream and call for his mother. "Just a few
miles more," I urged, rocking forward, as if the movement might
actually propel us ahead a little further.

"Goddamnit, Benny, we're not going to make it, are we?" The
copilot's voice was shaky and had risen an octave.

"We'll make it," I answered. I looked over with a reassuring smile. "I'm taking us down. There's a clearing up ahead. Call in our coordinates. Tell them to send in the cavalry."

While the copilot changed frequencies I added, "don't worry, I'll get you home to your wife." The copilot nodded, relieved. If Captain Benjamin Olstein said he was going to get you down in once piece, by God you knew he was going to get you down in once piece.

Controlling a helicopter was more than rules and mechanics, it was a feel—a special uncanny instinct between a man and his machine. But, only a few possessed it. Benny Olstein had a reputation as one of the best chopper pilots the Army had to offer. His flying record in Nam had already earned him the rank of Captain.

"Don't worry, Cap'n. Tomahawk and me will protect you just fine." It was the gunner. He'd finally ducked inside the cab, pulled on his helmet and was wearing a crazy-assed grin while patting his M-60 affectionately.

"Thanks Crazy Horse," I keyed. "We're gonna need your expertise."

"You can count on me, Cap'n."

I hope to fuck I can, I thought.

Then, as if Crazy Horse had just read my thoughts he whooped into the mouthpiece and shouted, "Goddamned right, mother-fuck-errrrrrrr! I will kick ass and take names! Fuck that, I ain't even takin' names, I'm just gonna kick ass!" And with that he fired off a few rounds for good measure.

The co-pilot rolled his eyes and shook his head. "Lord help us," he said nervously crossing himself.

Cold air rushed in from the cargo door. The blonde kid with part of his head missing hadn't moved since being thrown on board and I wondered if maybe he was already dead. The black soldier who seemed to be standing guard over him apparently thought otherwise and continued to talk to him and smooth his hair.

The soldier that had been flailing with his stumps was quiet now. Maybe the morphine had done its work, or maybe he'd just fallen unconscious . . .

. . . This hadn't been a part of the plan, I thought. I wasn't taking part in a dust off, or flying medivac. I was flying a UH-1D, equipped for nothing more than transporting soldiers—an air taxi to insert troops into the field—a slick. "Slick, it rhymes with dick," the flight instructors told the pilots. "Slide 'er in, get your load off, and get the fuck out."

But, something had gone terribly wrong.

The mission had started out okay. Everything had gone like clockwork. It was going to be an easy operation—and up to a point it was. Like items being checked off on a laundry list, everything went according to plan—well, almost.

0500: Wakeup. I emerged from a fitful sleep drenched in sweat. It was a warm Thursday morning at Camp Holloway, Pleiku, in the Central Highlands of the Republic of Vietnam.

Outside my hooch the steady beat of helicopter blades had lulled me to sleep and I fell in out of dreams as quickly as Hueys arrived and left for the field. Thankfully, the dreams made no sense. But, Nellie had appeared in some, so the emptiness of longing lingered.

I had plenty of time before my preflight briefing, so I trudged across base to the Air Force's mess tent where I knew the food would be better. I strolled with a sure, relaxed gait and took long strides as I smoked a cigarette and watched the sky grow light with the dawn. The dirt at Camp Holloway was a fine powder, a dark red talcum that stained my fatigue pants and turned the air musty.

When I reached the mess tent, I stepped over the sandbags and pushed the bug screen aside—ducking my head as I entered. It took a moment for my eyes to grow accustomed to the low light.

Eating breakfast in an Air Force mess tent was akin to dining out. They used blue canvas for their tents, blocking the sunlight and plunging the interior into deep shadows.

Throw in a few candles, some checkered table clothes, and put Jackie Gleason on the hi-fi and they could charge admission, I thought.

Breakfast was an orange malaria pill washed down with black coffee, creamed beef poured over hot rice and a side of soft scrambled eggs. There was plenty of fresh fruit that morning so I piled my plate high.

While I ate I reread Nellie's last letter. She was excited about my upcoming R & R and had just bought plane tickets.

You know, she wrote, *if I'd stuck with the ROTC Program instead of bailing out when we relocated to the Bay Area, I'd be flying standby for free. But, I don't care— I'd pay ten million bucks for these tickets—and even that would be a bargain!*

In less than a month we would rendezvous in Honolulu for ten days of sun and fun. Or is it fun and sun? She teased.

It would be the first time I had been outside of a combat zone in seven months. From the day I left her standing at the departure gate, I felt like I was walking bent over, leaning into tomorrow, reaching for that space in time when I could hold her in my arms again. "I ache for you," I wrote to her. "When I let myself think about you for too long my stomach feels like I haven't eaten for a month."

The end of August will be a good time, she wrote. *Classes don't start at Berkeley until September and I'm pretty sure I won't be on my period.*

A very important consideration, I noted. That girl was regular; hadn't missed predicting a start date in four and a half years. But, even if she was off and started early, it wouldn't make any difference.

I think about you every waking moment. Which makes it hard to study sometimes, but I wouldn't have it any other way.

There was a bad storm Tuesday last. A giant Maple tree that used to sit next to the backporch at Harold and Irma's fell over and crashed into their living room. Luckily it happened in the middle of the night and nobody was hurt.

Harold said it sounded like a freight train driving the house. Said he predicted it would happen someday and it did.

Mom called today and said Grandpa was back in the hospital. He just can't seem to stay away from those strawberries. Especially if they're smothered in cream and sugar. They'll be the death of him yet. Anyway, he's feeling poorly, but Mom says his spirits are good.

He showed Mom a letter he got from Jeffery.

From what I can gather it sounds like he's stationed over there with you somewhere. He's assigned to the Army Rangers. There was a picture enclosed. She said he was wearing a Green Beret—but would have passed right by him on the street and never recognized him. She said he'd changed—that he looked a lot different—older, I guess.

Yeah, I thought, Vietnam does have a way of aging you, alright. I examined the photo. Nellie was right, I wouldn't have recognized him if walked up and shook my hand.

Not to change the subject or anything, but I sure do love you—ya big lug, ya. I'm glad you're keeping up with your writing.

That last short story you sent me was the funniest thing I've ever read. Someday you'll be a famous author, Benny. I just know it.

When in the hell do you find time to write, anyway?

Oh, by the way—I love you, I keep coming back to that, don't I?

In your last letter you said you ran into an old high school friend of yours. Stanley? That must have blown your mind. I think you said you almost shit your pants. Or did he almost shit his—I forget. Which was it?

Love you for ever and ever until Hell freezes over, Your loving, beautiful and fabulously wealthy wife,

Nell

P.S. Blowjob
Just something for you to think about. —Bye

"Must be pretty good reading, Captain—looks like your food is getting cold."

"Speak of the devil," I said cheerfully, putting Nellie's letter away. "What brings you to this neck of the woods, Stan?"

"Hey, I work here, buddy," he said, tapping his shoulder patch. "37th Aerospace Rescue and Recovery Squadron. The super-dooper elite of the elite." He let his food tray clatter onto the table and took a seat across from me.

"Besides, in all modesty, a guy's gotta eat. You know, if the word gets out the Air Force has the best chow, all of this will disappear and you and I will be forced to eat C-rations like everyone else. Normally, we don't let Army grunts in here, but in your case, being as you're an officer and my best friend, we're making an extreme exception this morning."

"Oh, your Benevolent Majesty, your generosity knows no bounds." I bowed my head slightly. "I applaud your graciousness."

"Please—hold your applause until after the war. What were you reading there— another letter from Nell?"

I nodded and Stanley began shoveling food into his mouth like it was his last meal. "When in the hell does she find the time to study?" he asked between bites. "She must write you a letter every day."

"Just about."

"Do your folks know yet?" he asked. I shook my head and grinned. "Do hers?" I shook my head again. "You're both out of your fucking minds, you know."

"Maybe."

"No two ways about it. Both of you should be fitted for straight-jackets. Cousins don't marry each other."

"We did. So did Jerry Lee Lewis. He married his cousin, didn't he?"

"That's different."

"What's different about it?"

"Jerry Lee Lewis is a rock and roll legend. Legends don't go by the same rules; they make the rules. Besides," said Stanley shoveling in another mouthful, "I have it on good authority that he's probably hopelessly insane. Aren't you worried about your parents—how they might take it?"

"Not in the least. Actually, I could give a shit less."

"You, my friend, are in dire need an attitude adjustment," said Stanley, punctuating his words with his fork. "Look around you. Here we are in the heart of the mysterious East having the time of our lives and you're using profanity and disrespecting your parents. You should be ashamed of yourself . . . you fucking bastard."

I laughed. "You're right. I don't know what I was thinking of. I must be out of my mind."

"See? That's what I've been trying to tell you," said Stanley. "Do you really think the food is better in the Air Force's mess hall?"

"Most definitely."

"Indeed." Stanley ate thoughtfully for a moment. "I'll never forget what a great cook your mother is," he said, changing the subject abruptly. "Remember when you had me stay for dinner a couple of times? I never in my life tasted food so good."

"Doesn't your mother cook?"

"Sure she cooks, but only to stay alive. You know, steak and potatoes, meatloaf and potatoes, pork chops and potatoes, sometimes spaghetti. But she never went to the trouble your mother did, sprinkling lightly sautéed tiny mushrooms around a platter of roast beef topped with giant onion rings that only moments before were simmering in butter."

"Stop, you're making my mouth water."

"And those tossed salads. Whoever would have dreamed of smothering lettuce, tomatoes and cucumber in parmesan cheese? And her desserts were out of this world. I have never tasted a pecan pie like that before. What did she call it— Washington Nut? I swear, your mother could open her own restaurant if she wanted."

"She might have to, now."

"Why's that?"

"My parents are getting a divorce."

Stanley stopped eating. "Are you shittin' me?"

"Nope." I picked up a slice of green melon with my chopsticks. "Hey—not bad. I'm getting pretty good t this, aren't I?"

"How long were they married?"

"Twenty-four years, I think. Something like that."

"Jesus," said Stanley resuming his meal. "What a catastrophe. I just can't believe it. The whole world's gone completely nuts. Do you know why?"

"Why the world's gone completely nuts or why they're getting a divorce?"

"Either one. Take your pick."

"No, not really," I shrugged—but, I had a pretty good idea. Nothing was the same after sending Jeffery away. The subject never came up, but eyes seldom smiled and words became few and stilted. My sister, Holly, stayed up in her room, rarely venturing into the family fold again, and I spent most of my time as far away from home as possible—practically living at the public library until I finally moved away to college.

The house was cold even when the sun shined. The living room became unlivable; the furniture lost its luster and grew stark; the walls lost their color and dimmed drab. The phonograph sat silent; no one turned on a lamp to fill the emptiness and I avoided the room altogether. Gradually, without caring, our love dissipated and we all drifted apart.

"Just the same," Stanley shrugged, "they're going to find out sooner or later. What then?"

"What the fuck do I care? I'm over here and they're over there. Yesterday I took fifty NVA rounds. I was lucky the fuselage didn't blow up. You know Spengler, my wingman?"

Stanley nodded. "We've met."

"He watched tracers shoot up through the seat between his legs." Stanley had stopped to listen, his fork poised halfway between his plate and his mouth. "He turned to me with eyes as big as saucers and said, 'holy shit.' That's when his left ear got shot off. So, do you really think I give a shit what my parents might think?"

Stanley squirmed in his seat, embarrassed he'd brought the subject up in the first place. Benny was right—what the fuck difference did it make?

"Well, let's change the subject," Stan offered.

"Sounds good to me."

"Let's talk about Donna."

"Donna?"

"You know, the goddess with the big tits in science class. I still fantasize about her."

How strange, I thought, here I was seven thousand miles away from home and who should I be eating breakfast with, but my old high school chum, Stan—now Pararescue Staff Sergeant Stanley Richards. What were the odds of that happening? A million to one—a zillion to one?

As I watched Stanley wolf down large portions of rice, I wondered if their forgathering was somehow a part of a master plan, or just happenchance.

We'd met up two months earlier on an unusually windy day when the delta winds got angry and swept down from the highlands with terrifying force. In the jungle the trees shuddered, and the leaves slapped one another in a death rattle; the turbulence sent choppers flying into one another, tossing the ten-ton flying machines around like they were plastic toys.

I was flying that day, returning to Camp Holloway from a mission when the call came in. The Wing Commander, piloting the lead Huey of a squadron of fourteen, took the mike and the rest listened in:

"We've got a problem, Two-Zero. Wonder you could lend us a hand."

"That's what we're here for, Skipper, what's up?" The Wing Commander was an ex-navy pilot and called everyone "skipper."

"We've got a bird down."

"Where? What the situation?"

From the exchange, I learned that a scout ship had been shot down in the Central Highlands. The pilot was still alive and had radioed his position. The gunman was unconscious, his condition was uncertain.

"What about the Blue Team?" the Wing Commander asked. "Where are they?"

Blue Teams were quick response reinforcements that were usually deployed for just this sort of operation—rescuing the crew of a downed chopper. It was a reaction force made up of a lead Huey, two Cobra gunships and sometimes additional slicks for inserting infantry landing forces if they were needed.

"They're in trouble, Two-Zero."

"What sort of trouble?"

"Rockets. We've got one Cobra down—no survivors.

There was a pause.

"Marine ground forces were inserted, but were hit hard. We've got a lot of wounded. They were ambushed—walked right into the middle of an NVA Regiment."

Benny exchanged looks with his copilot, Lt. Spengler, who keyed the cockpit mike. "Holy shit," he said. "They're using the first downed chopper as bait."

There was never a question about responding. If another bird was in trouble, a rescue attempt was standard operating procedure.

"Alright, who's got wounded on board?" the Wing Commander asked his squadron. There were a number of responses. "Those of you with wounded stay on course. That leaves five of us." He read off the new headings. "We're about twenty minutes from contact. Let's do it."

The five camouflage-green UH-1D's banked hard to the north and throttled up." I was piloting one of the five.

The terrain was a patchwork quilt, a collaboration between nature and man. Neat squares of constantly tended rice paddies, bordered dense pockets of forest. Rubber plant farms with trees so perfectly aligned they must have been measured with a ruler stood like toy soldiers next to an untamed jungle. The sky was so blue and clear it was difficult to imagine the devastating carnage that was ripping the country apart below.

Ten minutes into the flight my hands were sweating so bad I had to wipe them dry on my pants. It was time to come to grips with my mortality and wonder if this was it; if the die had been caste; if these were the last few moments of my short sweet life. It was time to make peace with my Maker.

I prayed silently, sometimes in English, sometimes in Hebrew, but always with Grandpa Leo's voice ringing in my ears. And why not? It was Leo who'd taught me to pray in Hebrew, Christ taught me to pray in English . . .

. . . I skipped the last day of my junior year in high school, knowing there would be no work and it was really just a day for good-byes. But, that wasn't the real reason I didn't show up for classes.

In truth, I was having a tough time facing Donna after running out on her that day. She'd already corralled me once and we were both late for class.

The halls were vacant and she came out of nowhere, pushing me between a locker and the drinking fountain, forcing me against the wall.

"You've been a very naughty boy, Benny," she teased, smiling coyly. I felt myself break into a sweat. "And very naughty boys should be punished."

"Look, Donna," I managed.

"What do you suppose your punishment should be?" she asked, softly.

Maybe a little levity is what the moment called for. "You could always beat me to death with your breasts," I suggested.

She suppressed the urge to laugh, then whispered, "I've got a better idea." Her hand slipped to my crotch.

"Aren't you two supposed to be in class?" To my relief, it was the hall monitor.

Thank God for hall monitors, I sighed. So, instead of going to school, I spent the morning in the basement stacks at the University of Washington reading Alexander Dumas and Henry Miller. A strange combination, but when I stepped outside I felt refreshed and the misting rain felt good.

I kept my umbrella folded and walked home, following a path through the arboretum. I let the rain soak me until my hair was plastered to my head and rivers of cold ran down my back.

The front door was locked and the house was quiet except for the grandfather clock that ticked in the foyer and our dog Cindy, who hopped up and down with her nails clicking on the floor until I finally knelt and gave her some attention.

My mother had gone shopping with my Aunt Fran and Grandma Olstein, leaving behind a note saying they'd all be having dinner downtown at Bob's Chili Parlor and to meet them there at six. "Be sure to dress nice," the note ended.

My mother left an open pack of cigarettes on the kitchen table and I helped myself, storing two in my shirt pocket and slipping a third between my lips. I grabbed a handful of wooden matches on my way to my father's study.

At this time of day, with the window shades drawn, the study would be steeped in a warm glow of apricot. I planned to leave the lights off and settle into my father's chair to bask in the warmth of the hundreds of volumes of books.

I could already imagine myself settling back and putting my feet up on his desk as I struck a match on the door and pushed it open. I knew my mother would have had a fit had she seen me do that and grinned at my defiance, fancying myself a true rebel. I was just lighting up and sauntering into the room when my Grandfather Leo turned to face me.

"Benny, do you mind?"

I gasped and inhaled too deeply. "Grandfather," I squeaked as the smoke painfully expanded in my lungs.

"I'm almost done," said Leo. His voice was soft and amused. A white tallith covered his shoulders and phylacteries were strapped to his forehead and left arm. He was holding a little black book.

"Sure, sure," I whispered, barely able to find my voice as I backed out and closed the door. I coughed my way back to the kitchen and collapsed into a chair.

When I recovered, I flushed with embarrassment. I'd been caught smoking. What was worse, I'd struck a match on my mother's mahogany door—the one she'd personally stripped and sanded and given seven agonizing coats of shellac that stunk up the house for nearly a month. Maybe what embarrassed me the most was my grandfather's religious garb. Leo was an Orthodox Jew, and I'd interrupted him during a moment that was obviously a private one and none of my business.

I was still pondering this when Leo walked quietly into the kitchen, the tallish draped over his arm.

"Is there someplace where we can get a malted?"

"You mean like a milkshake?"

"Milkshake, malted, whatever," said Leo.

"But, it's raining out."

"So I can't have a malted when it's raining?"

I laughed at myself. "It's just that I never have one unless it's sunny."

"In that case you probably haven't had very many malteds since you moved here."

"Come to think of it, no, I haven't."

"So?"

"So, there's a drugstore about three blocks from here with a soda fountain where you can get one."

"Good. Come with me. You like malteds?"

"I like chocolate milkshakes.

"Let's go," he said and headed for the door. "Unless you'd rather wait for a sunny day. We could all be dead by then."

He stood on the porch and put his umbrella up while I locked the door behind us. He was still wearing his yarmulke. I almost said something, but decided I might offend my grandfather by pointing it out, so I didn't mention it.

"Here," said Leo, handing me an envelope as we headed down the hill sharing the umbrella. "You got a letter."

It was from Nellie. I unconsciously sniffed the envelope then shoved it into my pocket.

"I think that's the second letter this week, if I'm not mistaken."

I smiled. "You're not mistaken."

"Yes, the last one smelled like that too. I don't usually smell letters. If I get a letter, say a bill from the telephone company, I never bother to smell it—I just open it. But hers. . . ." His voice trailed off and he let the thought hang.

Oh great, here it comes, I thought. What if he suspects? If he does, should I make excuses or just come right out with it? That's right, goddamnit, make a federal case out of it. I'm in love with my cousin, alright, are you satisfied? We write each other a lot. We've kissed; we've even . . .

"So, you're home early from school."

"What?" I wasn't expecting the sudden change in subject.

"School," Leo repeated.

"Oh yeah. I guess I came home early."

"They're lucky you bother to attend at all. You're wasting your time and theirs. Go to college, Benny. Your age doesn't matter."

"Maybe I will."

"Maybe, shmaybe. I know you will. I can tell. Anybody who teaches himself Italian because he's bored belongs in college."

I shrugged. It was just an exercise that had gotten out of hand.

"Ever wonder why you chose to learn Italian?"

"Actually it began when I decided I wanted to read Boccacio. I wasn't happy with any of the translations of 'The Decameron' I came across."

"See? That's what I mean. Why would anyone want to read that book? It's not even required reading in college."

"How do you know?"

"I know these things. So, why teach yourself Italian?"

"Because 'The Decameron' was written in Italian?" I asked, grinning.

"That's not why, Mr. Comedian. You see, you think you know why," said Leo tapping his temple, "and you don't know why."

"Okay, you tell me then."

"Because it's in your blood; you couldn't help yourself. You have ancestry from Italy."

"I do?"

"Sure, you do. Didn't your father tell you about your heritage?"

"A little. I know you came from Russia."

"So you think I'm Russian? I speak Russian, but I'm not Russian, I'm Dutch. My father came from Holland to Russia when he was just a little boy."

"So where's the Italian come in?"

"Your grandmother. Sure. Rappaport. That was her mother's maiden name. And her father was a Rabbi in Northern Italy. You've got Italian blood in you—a lot of it."

"A Rabbi?" I said, pondering. I had ancestors that were holy men?

"Sure, sure, a Rabbi. You didn't know you were so closely related to God?"

"Is that what you call it?"/"

"What would you call it?"

"Well," I said, "I'm not even sure I accept the concept of God."

"Benny, Benny, Benny," Leo chuckled. "You're much too intelligent to be an atheist. You believe; you just don't know you believe. Do you believe in destiny?"

I thought about Nellie then answered, "I think so."

"Then you believe in God."

"How so?"

"Everything has a reason, Benny. You think you're here because of some combination of soup made from salt water, lighting and exploding volcanoes a million billion years ago? That's ridiculous. Your existence is not an accident. Watch yourself crossing the street here," he said, holding his arm up like a roadblock. "The roads are slick and nobody knows how to drive in this town." A truck roared through the intersection and honked its horn for no apparent reason. "See what I mean?"

"What about evolution?"

"What about it? You think maybe you've got a relative that was monkey?"

"It's possible."

"There are no Jewish monkeys. It wouldn't be kosher. God made us special. You know why there's a missing link?" I shook my head. "Because there was never a link to begin with, that's why. You can't change lead into gold, Benny. There's no link. Does it ever stop raining here?"

"Sometimes."

"Well, let me know when it happens. Even if it's in the middle of the night—wake me. I want to see it. I'll pinch myself so I can remember the moment. I should have worn my rubbers," he said stepping over a puddle. "My shoes are getting ruined."

"So you think all the scientists are full of shit?"

"Well, I wouldn't exactly put it that way, Benny."

"I'm sorry," I said. "It just slipped out." Leo looked at me sidelong and we both laughed at the pun.

"Listen," said Leo, "man's always looking for a scientific explanation for everything." Leo protectively took ahold of my arm as we crossed the street. "Some things are sitting right in front of their nose—staring them in the face and they don't even see it. Too many things are too perfect for it to be chance. Here—hold this, my arm's getting tired," he said, handing me the umbrella.

"You speak English so well, no one would even suspect it's your second language."

"Second language? More like fourth or fifth. But, thank you for the complement. Complements are hard to come by."

"Fourth or fifth?"

"Sure, sure. I speak Russian because I was born in Russia. My father, he didn't speak Russian so good—enough to get by—but he was Dutch—he spoke Dutch, so I learned Dutch. My mother, she was German, she didn't speak Russian so good either. So, I learned German. Then of course there's Yiddish—everyone spoke that. No matter where you're from, that's the language that binds us together. And, in shul you had to learn Hebrew or you wouldn't know what you were praying about. They taught us a little English in grade school—but, I didn't really pick it up until I moved here about forty years ago. So how many languages is that? I lost count."

"Six."

"Six—not bad for an old man who never went to high school."

"Do you believe in destiny?" I asked.

"Of course. Let me tell you something. I don't drink and I don't smoke. But, my father, your great grandfather, thank God he did. If he hadn't, you wouldn't be alive today."

Leo stopped in his tracks "Look at that," he said, bending over to examine the sidewalk. "Slugs. Everywhere you go in this town there's slugs. The place is crawling with them. Watch your step," he said, taking my arm as if to guide me. "You could slip on one and break your neck." We gingerly stepped over the slithering mollusks.

"If it hadn't been for your great grandfather's vices, you wouldn't exist. The woman you'll someday marry would never meet you—she would meet someone else and have his babies. You're grandchildren wouldn't be born—hell, your father wouldn't have been born for that matter."

"How do you figure?"

"If your great grandfather hadn't died when he did, I wouldn't have had to go to work and become an upholsterer. Maybe I would've finished school. Who knows? Maybe gone on to college, become a doctor or a lawyer—God forbid, or something else; maybe I never would've left Russia—never met your grandmother. Only God knows. The point is, other paths would've been taken, the world would have been a different place."

"That sounds more like cause and effect than destiny."

"Destiny, cause and effect, what's the difference? Who's to say they're not the same thing?"

"Where did you meet her anyway?"

"Your grandmother? In New York. I hadn't been in this country two weeks when I saw her for the first time. She was sitting outside on the front steps of an apartment house in Harlem. She was just a little girl. Twelve or thirteen, I think. I was twenty-three, but I knew she was the one, so I waited. I introduced myself to her parents, told them of my intentions and when the time came, we got married."

"It was an arranged marriage?"

"Things were different back then."

"Harlem? I thought Harlem was the black ghetto."

"Back then it was mostly white—Jews and Italians. Eventually, we moved out— moved up—and the Negroes moved in. Anyway, we're getting away from the subject. We were talking about my father—your great grandfather. His name was Henry, like your father."

"I didn't know that."

"Your father never told you?" I shook my head. "Well, it's not important. What's in a name? When they changed my name to Olstein did I balk? No. I was just happy to be here—in a free country. They could call me whatever they wanted, it didn't make any difference. I was still me, and that's all that mattered."

"What was my great grandfather like?"

"Like?" Leo shook his head as the images of his life in Russia swarmed in. "He was a big man—side-chops, you know, lots of whiskers where sideburns go. He was a good Jew. He wasn't devout, but he kept the Sabbath—most of the time. And the one time that I know of when he didn't, was the most tragic mistake of his life."

"Is that when he died?"

"Soon afterward," said Leo shaking his head.

"What kind of work did he do?"

"He was an artist. A window dresser. He traveled the city of Odessa on foot offering his services to shops and stores to decorate their display windows. He was very good at it, too. He was able to support a wife and five children doing it.

"And his father was an artist, too—in Amsterdam. He painted portraits and even taught art for a while. So, not only do you have Italian blood, and holy blood and Jewish blood in you, but you also have the blood of artists running through your veins." He laughed and patted Benny's back affectionately. "You had no idea you were so important, did you? I think I'll buy a newspaper," he said, stopping and feeding coins into a machine.

"Writing's an art. I know you write. You've been writing since you were a little boy. You used to send me plays and poems that you'd written when you were five and six. Remember that? I'll bet you still write."

I shrugged. "I can't help myself. I feel better if I have some sort of writing project going on."

"See? That's what I mean." He briefly studied the headlines then folded the paper and tucked it under his arm. "You can't escape your blood, Benny. It's in you. You're probably religious and you don't even know it."

I laughed.

"You laugh—what's so funny?"

"How could I be religious and not know it?"

"Simple. I'll tell you in one word—Saint Paul."

"That's two words. How would you know about St. Paul?"

"I read. What—because I'm a Jew I don't know about St. Paul? Everybody knows about St. Paul. First he doesn't believe—and then wham!—before you know it, he's the first Christian. Explain that."

"He had a vision."

"Vision, shmision—he just became aware of his beliefs, that's all. He always believed, he just didn't realize he believed until one day it dawned on him. How much further is this soda fountain-drugstore of yours?"

"Another block or so."

"Or so? What does that mean? Is it another block or not?"

"A block and. Half. You were explaining your theory of destiny."

"Theory? What theory? I was talking facts."

"How did my great grandfather's drinking affect my life?"

"If he hadn't got drunk that night you wouldn't be alive today. He was only 38 years old, you know. It happened in the fall. I remember the time well. Most of the trees had already dropped their leaves and at night I would need extra blankets. The window to my bedroom closed all the way, but the cold air always figured out a way to sneak through the cracks in the wood . . ."

. . . His work day had ended. Thank God for that. He'd managed to complete three windows. A good day. Good money. Enough for food and rent and a little left over for himself.

It was getting dark. The Sabbath was beginning and it was time to head home.. His wife would already have baked a fresh challah and placed it on the table.

But, Henry had been nipping on and off all day and by now he was feeling too good to go home, just yet. It wasn't only the alcohol that warmed his blood, though it sure had something to do with it, but the warmth of achievement. It had been a very satisfying day.

Three windows! Ah, and that last one had been a beaut. A real looker. He stood on the sidewalk outside the shop after he'd finished and stared at it for nearly half an hour. He even walked across the street to look at it from a distance.

Oh, she was a beauty all right. Henry took a swig and sighed heavy. The gas lamps hissed overhead and the sky was clear and growing dark, but the light from the lamps made the stars impossible to see.

In spite of the brisk air, Henry was warm and never bothered to button his wool coat as he made his way for the harbor. When he felt good, he whistled—and tonight, he whistled.

By the time he reached the piers, the sky was black and the wind had died to a whisper. Many of the whores he'd passed had removed their jackets to display their wares.

Most of the male passersby ignored them, others made snide remarks—but, not Henry. He never held anything against them and secretly thought them daring if not downright brave to do what they did.

When asked if he was interested he politely declined, refusing to compromise his fidelity. But, he always had a kind word or a complement for the girl, even offering to pay for a meal if one looked particularly down and out.

Over the years, since he frequented the bars near the harbor often, he was acknowledged with a nod or a smile—some even called his name when he passed. Those who didn't know Henry assumed his familiarity with the whores was for unsavory reasons and shook their heads in disgust.

On that night, he chose The Kalinka, a bar whose doors faced the waterfront. He promised himself he wouldn't stay long—that he'd head for home soon after a couple of drinks, and pushed his way inside. A few of the patrons knew him and hailed or slapped his back as he settled onto a stool and ordered a beer.

"Ach!" came the voice of a bearded fellow two tools away. "Your money's no good. Let me buy you your first beer." And with that he slapped the coins onto the bar.

Henry smiled. "I'm afraid I don't know you, friend."

"No, you don't. And, you're not likely to see me again after tonight either." He moved a stool closer. "I'm off The Klondike, a freighter anchored at pier forty. Ship's mate," he said proudly and extended his hand to shake.

"The Klondike," he repeated as his mug was served. "Is that English?"

"English? Good God, no. It's American. Out of Boston."

"Ah," Henry nodded knowingly and took his first sip. "The United State of America."

"That's right!" said the stranger a little too loudly. He took a healthy swig and brought his mug down hard on the bar. He smacked his lips with a gratification that comes from over indulgence and smiled crookedly. His face and particularly his nose had lost most of its feeling. He breathed a heavy sigh before speaking again. "Ever been there?"

"To the United States? No," Henry chuckled and shook his head. "Never have. Never been anywhere, really. Have you?"

"Sure, I have. I live there. I've been a citizen for three years now. Smartest move I ever made."

"Why do you say that?"

"Because it's the truth. It's the greatest country on Earth. A noble experiment in freedom that worked, by God. Drink up and I'll buy you another." He threw back his head and poured the contents of his mug down his throat.

Two beers soon became three, then four and before long Henry lost count. When he tried to figure it out by counting the empty mugs the mathematics became a formidable task. The next time he blinked he found himself outside in the alley behind the bar relieving himself with his new found friend who was laughing and peeing all over the side of the building.

"I should be getting home," Henry mumbled.

"Yes, and so should I," his friend agreed. "What time is it?"

"I have no idea."

"Neither do I. Nor do I give a rat's ass."

Henry struggled to pull his watch from his pocket while his friend hummed and swiped a yellow stream back and forth over the bricks. "I don't think I've ever peed so much in my life."

"It's ten after nine," said Henry closing the cover and shoving the watch back into his pocket.

"Ten after nine! Jesus Christ," shouted his friend as he stuffed his penis back in his pants. "The boat leaves in twenty minutes, come on." He continued to relieve himself, wetting his pants as he grabbed Henry's arm. Before Henry realized it, he was running alongside his American friend.

Now that the hour was late, the whores were out in full force. The sight of the two men racing by was a welcome distraction from a night that had turned chilly and slowed business. "What's the hurry, boys?" Henry grinned in response and tipped his hat as he hurried past.

On more than one occasion a lady of the night would pull down her bodice and expose herself. "What do you think, Henry?"

"Very nice," Henry would reply, smiling sheepishly as he was pulled along by his friend.

"No charge for you tonight, Henry. I owe you," one called.

"No, no. You owe me nothing," Henry called back. "I only do what any man would do."

"That's why you think," came a retort, knowing Henry wouldn't hear.

"You seem to be a very popular man," said his friend, breathing hard, but keeping a steady pace.

"I can't imagine why," said Henry, a little embarrassed by what his friend might wrongly assume.

"Can you help me out, Henry?" a small woman pleaded. "I haven't eaten all day." Henry stopped suddenly and pulled his arm free from his anxious companion.

"Do you have a place to stay tonight?" Henry asked, reaching into his pocket and pulling out a handful of change.

"Oh sure, I can stay with my sister. She'll put up with me for the night—I'm sure off it."

She was short and not at all unpleasant to look at. Her dress was old and worn and her vest a little too tight, but it only served to accentuate her waist making her all the more attractive. She tried to smile, but her eyes couldn't mask her uneasiness. She was frail and needed a bath.

Henry eyed her warily and pulled some bills from his pocket. "Here," he said pushing the money into her hand. "You get yourself a good meal and a clean bed tonight. You hear me?"

"Oh, Henry," she said, kissing his hand. "You're such a dear."

Henry blushed and quickly withdrew his hand. "Well, never mind that," he said, flustered.

"Come on," said his friend, pulling him away. "There's no time for that now. You can get yourself a woman later."

Henry stumbled a little then recovered as they picked up the pace once more. "God bless you, Henry," she called after him. He turned slightly and attempted to wave back, but had to suddenly grab for his hat as it attempted to fly from his head.

"We're almost there, his companion panted.

"Where?"

"Why, The Klondike, of course." Just a head an old freighter was moored, dipping and rising slightly out of the water. Buoy bells clanged in the midst, the ship groaned and nudged the quay. Lanterns were burning on deck, hands were preparing to cast off.

"You just let me do the talking," his friend said as they hurried up the boarding plank.

"I have to sit down," said Henry as they stepped onboard. He was out of breath and felt he was going to collapse. His friend led him astern and eased him onto the deck, propping him against a stack of crates.

"You'll be all right here. You just rest—I'll take care of everything. I'll check back with you later." His friend disappeared and Henry closed his eyes.

What on earth was that man talking about? Henry wondered as he was swallowed by a gentle hum that set his mind adrift. Take care of everything? What's there to take care of? His body floated, rising and falling with the motion of the ship. The fog from his breath told him it was cold, but that was his only clue; he felt hot. It's this damned wool coat, he thought, shrugging it off. Sweat ran into his eyes, stinging.

He'd been unusually warm all day and wondered if maybe he was coming down with the flu or something worse. Earlier in the day several people mentioned they'd felt under the weather and complained of aches and pains.

The gentle rocking of the ship was soothing, touching a distant memory. A mother, many years deceased, reached out from the past and cradled her infant son, rocking. And, like the waves that lapped the keel, she whispered lullabies long since forgotten.

He didn't know how long he slept, but he awoke with a start and an ache in his back from leaning against a rope.

"Ah, at last you're awake," his friend said grinning widely and patting his shoulder. "You're a lucky man!"

"Tell that to my head," said Henry struggling to his feet.

His new friend helped him. "There, you see? Odessa looks beautiful at night, doesn't it?"

Henry rubbed his eyes and peered into the dark.

"But, that's nothing. Wait until you see New York. Now, that's a sight to behold!"

Henry tried to clear his head, but the booze had done its work, and now his body was busy converting alcohol into formaldehyde. The slightest movement was misery. His head had become a fragile egg caught in a vise, his skull collapsing under the pressure, cracking open, his brains painfully dripping down his temples.

He stumbled to the railing and took a deep breath, filling his lungs with the icy air, hoping it would help. When he looked down he saw nothing but black. The boat dipped and made him queasy. When he fixed his eye on the horizon he was shocked by what he saw. "My God," he whispered. "We're moving."

"Of course we're moving. We weighed anchor and cast off some time ago. You missed it. I was going to wake you, but you were resting so peacefully I naturally assumed . . ."

"My God!"

"You must be very excited. You're going to America, my friend. Think of it."

"Are you crazy?" Henry shouted. "Turn this goddamned boat around—I'm not going anywhere."

At first, his companion thought Henry was joking, but the look of panic on his face told him otherwise.

"I can't turn this boat around," he explained, his voice wavering. "Nobody can turn this boat around."

Henry grabbed the stranger by the collar and brought his face so close their noses touched. "You've got two minutes to turn this goddamned boat around, or I'm going to strangle you and toss you overboard."

"You don't understand. I don't have the authority to do anything," he pleaded. "I thought you wa . . . wanted to go to America," he stammered.

"I don't want to go anywhere. I have a wife and five children. I can't leave them. I would never leave them! They're my life, you idiot!"

Then suddenly, before he could be stopped, Henry swung his legs over the railing and jumped.

Then, nothing.

His companion held his breath, struck by the stillness. Finally, he heard the splash as Henry hit the water. When he peered over the side, the lights of Odessa sparkled off the waves, bouncing like diamonds. Only a faint outline of Henry could be seen splashing and treading water below.

"Are you all right?" he called.

"Yes," Henry answered angrily. "I'll be fine."

He watched the ship pull away while his companion leaned over the railing, searching in vain, scanning the deep. The water churned, swirling about him like India ink. "So, this is why they call it the Black Sea," he mumbled, as he found his bearings.

With a little maneuvering he managed to kick off his boots and start for shore. That's when it dawned on him that he'd left his wool coat aboard ship. He groaned. All of his money was in that coat. "Stupid, stupid, stupid." He cursed himself and thrashed the water.

The swim was tiring and the water was cold. It wasn't long before he realized that losing his coat had been a blessing. He may have lost his money, but if he'd plunged into the water still wearing it, the heavy wool would have dragged him under.

Sometimes, overwhelmed with fatigue, he rolled onto his back and floated. When he did, he could see the ship. The Klondike was far in the distance now and getting smaller, steaming for Istanbul and the Mediterranean . . ..

. . . "And so," Leo stopped in mid-sentence and took a sip from his straw, "that's what happened." He was sitting next to me at the counter, a chocolate malted before him.

"That's it?"

"What else do you want know?"

"Did he drowned?"

"Drowned? No," Leo scoffed. "He was a good swimmer—a strong swimmer. He made it to shore okay."

"So, what happened next?"

"Next? Next he got home somehow—probably walked. He was soaking wet, of course. The next morning, my little brother, David, found him passed out on the front steps. He was in his stocking feet. Mother put him to bed. I went in to see him —he was sleeping. His forehead was very hot—he had a fever.

"The next day the fever was worse so mother sent for the doctor. But, there wasn't anything they could do. He just got worse and about a week later he was dead. Pneumonia." Leo stirred his malted, lost in thought.

I never cared for straws and drank my shake straight from the glass. "How old were you, Grandpa?"

"I was thirteen. I'd just been mitzvahed a month before. It was very hard for a boy so young to say Kaddish for his father. Jewish Law said I had to recite it everyday for a year. Which I did. I think I must have cried every time. I only say it once a year now, on Yahrzeit—that's what we call the anniversary of a death.

"Anyway, I did what I had to do. As the man of the house now, I knew my duties. I quit school and went to work. We had to eat. I became an apprentice upholster, which brought in enough for the family to get by and the rest is history."

"Is that why you don't drink?"

"I'm not sure. Maybe. No—that's not why. My little brother, David, is the reason I don't drink, I guess."

"Why—what happened?"

"One day—a few months after burying my father I came home from work and caught David playing in the street. He was throwing a ball up in the air and catching it. I said David you know better than that—it's not safe. Get out of the street. He said, 'okay, bossypants,' and I left him there. I went upstairs to our apartment and was putting my tools away when I heard this terrible noise—somebody screamed I think, or maybe it was the horses, I don't know.

"I hurried back down the stairs; I knew something was wrong. David was already dead by the time I reached him. He'd been run over by a beer wagon. Killed him instantly," he said, shaking his head. "He was only seven. I think losing David was harder than losing my father. David was special; so handsome, so bright. He looked up to me—his big brother. Said he wanted to grow up to be just like me." Leo absently stirred the remainder of his malted with a long spoon, quiet now. When he saw me check my watch he asked, "what time is it?"

"After three. We should be heading back pretty soon. I have to get ready—take a shower."

"Me, too."

"Dinner's at six."

"You don't sound very excited about it."

I shrugged; I wasn't. I'd eaten dinner at Bob's Chili Parlor so many times I'd lost count. For a long time it had been exciting eating at a cocktail lounge. The low lights, the smell of mixed drinks, the pretty barmaids with their short red skirts and black fishnet stockings and low scooped blouses.

"I went for a walk today," said Leo. "I found a nice little Russian restaurant just off Broadway. It's small but very respectable inside—I peeked in the window. They weren't open yet, I guess they only serve dinner. They might serve good food, who knows?"

"I don't think I've ever had Russian food."

"Why don't you skip Bob's Whatever-It's-Called and have dinner with me tonight?" I hesitated. "Just the two of us—you and me. You could call and leave a message at Bob's. Better yet, I could call your father at his office and let him know. Sure. It'll be fun. Sabbath begins at sundown, it'll be a nice way to end the week."

"Alright," I said. "Why not?"

"Good," said Leo finishing off his malted and pushing his glass away. "Then maybe after dinner we'll go to Temple together."

"Temple?" I was aghast.

"Sure, sure. I'll teach you how to pray, how to carry on a conversation with God."

"To services?" I asked, still trying to come to grips with the idea. "Oh, I don't know."

"Who knows? You might even have a good time."

"I don't think so, Grandpa," I said, shaking my head.

"Why not? Are you afraid?"

"I'm not afraid. It's not that."

"You think maybe you won't be welcome?"

"It's just that, I don't know the first thing about Jewish services. I wouldn't know what to do from one moment to the next."

"Nobody does the first time. I'll teach you. It's easy."

"I'd be embarrassed."

"Embarrassed? What's to be embarrassed about? The Jew who should be embarrassed is the one who doesn't show up at Temple."

"I'd be too uncomfortable."

Leo was silent for a moment. Then he looked at me. "Uncomfortable?" I tried to smile, but had a tough time doing it.

"Let me tell you something about uncomfortable. Before the last war, you had nearly thirty aunts and uncles and cousins. They were all rounded up by the Nazis and put in concentration camps because they were Jewish. You want to talk about uncomfortable? Now, *they* were uncomfortable."

I squirmed in my seat. Upsetting my grandfather felt worse than having my parents mad at me.

"And, I'll tell you something else. Not one of them made it out alive. You have no relatives in Europe anymore. Not one. You're the last of the line, Benny. The very last. If you don't have any children, it ends with you. Where's my check?" He snatched it from the counter. "Uncomfortable," he said, as if the word left a bad taste in his mouth. "Let's go," he snapped, sliding off his stool.

Outside, I took my grandfather's arm, hoping the affection would soften him. But, Leo was silent and refused to look at me. "You know, Grandpa, I've never been inside a church before. Never in my whole life."

"So?" Leo asked, finally looking at me again. "It's not your fault, Benny. I blame your father. I even blame your mother. Some religion, even if it's not Jewish is better than none. So . . ." he said, shrugging and trying to smile.

"So, this is really awkward for me, but maybe it's time—maybe going to Temple with you wouldn't be such a bad idea."

Leo laughed and patted me on the back. "You sound like a Jew already."

When we walked home together, Leo had a new spring in his step. "We'll have a good time," he said. "I'll teach you everything you need to know—it's very easy. You have a yarmulke?"

I nodded. "You sent me one for Christmas about a hundred years ago. I still have it."

Leo chuckled and patted my back. "You're a good boy, Benny. I send it to you for Chanukuh—your parents must have given to you for Christmas. It doesn't make any difference." Outside it was still raining, but neither of us paid it any mind. I held the umbrella and Leo held onto me.

That night, at dinner, the restaurant was quiet, with only the tinkling of glass and silverware apparent. The patrons spoke in hushed tones and the formally attired waiters padded softly on dark carpets. They wore white aprons over their trousers and carried towels neatly folded over their arms. On each table, a candle stuffed into the neck of an old wine bottle flickered while dribbling wax down its shaft.

"Why do we wear a yarmulke?"

Leo smiled inwardly at my use of the word we. "Remember what it felt like when you were a little boy and your father put his hand on your head to show affection?" I nodded. "It's the same thing, only now it's your Creator doing it and the yarmulke helps suggest that."

"And the prayer shawl?"

"It's called a tallith. The same reason, only now, God is putting his arm around your shoulder and pulling you close."

"Excuse me,, sir." It was the waiter. "Will you be ordering dessert?"

Leo looked at his watch. "We still have time if you want something," he said, but, I declined. All this God stuff was making me uncomfortable and there were moments during the meal when the food didn't want to go down—when it would expand in my throat then take its own sweet time dropping down to my stomach where it sat like a rock in a pond.

"Temple Beth Shalom," Leo told the cab driver and we settled back in our seats. On our ride over, Leo sensed my uneasiness and tried to comfort me. "Let me tell you a little about your history, Benny."

"More about the Olsteins?"

"No, I mean about the Jews. It's a long story, and we don't have a lot of time, so I'll sum it up for you in just a few words. We were the first."

I waited for more, but Leo seemed satisfied, folded his arms and looked out the window.

"The first? The first what?"

"The first everything." I was bewildered. "Who was the first civilization to have an alphabet—the Jews. Who was the first to believe in only one God—the Jews. Who do you think financed Christopher Columbus's voyage—the Jews. That's all you really need to know—the rest of our history can be summed up with name-dropping. We changed the course of history. We plotted the course of mankind."

"Maybe you should enlighten me."

"Abraham," said Leo, counting on his fingers. "He was the first Jew. Moses—he wrote the first five books of the Bible. Jesus Christ—probably the most famous Jew of them all. He taught them all how to pray, both Jews on non-Jews. Then there was Nostradamus, Marx, Mendelssohn, Chagall, Freud, Calvin Klein, Ralph Lauren, Levi Strauss, Artie Shaw, Benny Goodman, Houdini, Gershwin, Irving Berlin, Kafka, Proust, Ayn Rand, and Einstein—just to name a few. I think that just about covers it."

"You forgot Jerry Lewis."

"And, Jerry Lewis—how could I forget Jerry Lewis?" said Leo slapping his forehead. "And, Groucho Marx."

"And the Three Stooges."

"And Jack Benny,.."

"And Marilyn Monroe."

"She converted, but she's so beautiful we'll count her anyway. But, there's Elizabeth Taylor."

"I think she converted, too."

"What the hell do I care? A striking woman. I've been in love with her since National Velvet. It's just another sign of her good judgement and phenomenal intelligence." We both laughed. "The list goes on forever."

We were both chuckling when Leo felt Benny relax. "So, you see, there's nothing to be nervous about," he said. "You're in good company."

"I'm not nervous."

"Well, Benny, there'd be something wrong with you if you weren't a little nervous. This will be your first time in the House of God."

I caught myself chewing on my fingers and realized my grandfather was right, "Listen to me. There's nothing to be afraid of. God is not going to yell at you." Leo lowered his voice and spoke sternly. "Where the hell have you been, Benny? I've been waiting almost sixteen year for you to show up. What do ya think—I've got all the time in the world?"

"I guess I just feel a little awkward," I said.

"That's okay. That's normal. I've been going to Temple all my life and sometimes I still feel a little awkward. Just remember you've got a Friend with you every step of the way."

I smiled. I guess I never really thought of you as a friend, Grandpa."

"I wasn't talking about me." . . .

. . . I was jolted back to reality by the Wing Commander's voice crackling over the headphones. He was sitting behind the controls of a Huey UH-1D following the lead helicopter that was now reducing its speed. He and the pilots of the other four choppers were on a rescue mission.

"Alright, gentlemen, I do see smoke at two o'clock. Let.s hold our positions right here." The lead helicopter stopped and hovered. The trailing Hueys spread out in formation behind her. "We'll keep our distance until we get fix on the enemy."

"Hurry up and wait," I mumbled.

"Just remember gentlemen," came the Commander's voice again, "we leave no American behind."

"Goddamned right, sir." It was my Peter Pilot. He was so gung-ho sometimes he couldn't control himself.

"Who said that?"

"Lieutenant Spengler, sir."

There was a pause, then: "I like a man who agrees with me, lieutenant."

Spengler winked at me, visibly pleased with himself. I liked Spengler—he was a good co-pilot and did his job well, but sometimes the brown-nosing irked me to no end.

The squadron of choppers kept their distance while coordinates and the location of ground personnel were determined. It wasn't long before the Commander called in an air strike. I switched frequencies and listened. Soon two F-4's radioed they were closing in.

I waited. The cab's interior would have been cooler had we been moving, but just sitting there, hovering, the air became stagnant and the sweat ran in rivulets. I felt my back grow wet and when I leaned forward my shirt stuck to my skin. When I peeled it away the dampness was unexpectedly cool.

Very soon, winged silver bullets would streak across the sky, the roar of their engines reaching my ears long after the missiles dropped away and slammed into the enemy in a cacophony of murderous destruction. But, the minutes ticked by and nothing happened. The waiting was agony. The F-4's were still out of sight and inside the chopper the only sound was the constant clapping of the blades overhead.

"All hell's gonna break loose," said Spengler searching the horizon. "Any minute now."

And though he'd just predicted it, Spengler jumped as the hills in the distance exploded. Flames and black smoke billowed and rolled skyward.

"Yeah!" one of the gunners shouted.

"Now that's more like it," Spengler keyed, his eye wild with excitement."

A moment later another volley of bombs hit. Trees were ripped from the ground, roots and all. Some exploded like firecrackers, others burst into flame like matchsticks. This time the roar of jets was plainly heard as the F-4's thundered past.

"There they are! Goddamn what I wouldn't give to pilot one those fuckers." Spengler shouted over the din.

The fighters made a third and final run, dropping another volley of bombs—this time throwing a little napalm in for good measure.

"Nothing could live through that," said Spengler, staring at he aftermath.

I knew better. True, anything above ground would have perished, but the North Vietnamese Army was notorious for their tunnels and the hills were probably riddled with them. If they reach the tunnels in time, they might survive unscathed.

But even knowing that, none of it seemed real anyway. It was like watching a movie. The hills were far in the distance and the bright yellow-orange balls of rolling fire and spewing funnels of black smoke were fixed against the backdrop of a sky so blue they could have been on a Hollywood set.

"Steady, Gentlemen," came the Commander's voice. "I just received word that Pararescue is on the ground. They'll let us know when to move in."

I sighed and studied the ground to take my mind off the delay. Waiting made me nervous. "I don't like hovering like this," I said. "It's not safe. We're sitting ducks up here. Stay alert," I hollered, twisting in my seat to make eye contact with the gunners. "Keep your eyes peeled—the ass you save may be your own."

Crazy Horse was one of the gunners that day. He'd been chewing the same wad of Bazooka bubble gum since takeoff. He mopped his brow, winked and gave a thumbs up. "Fuckin' a-right, Cap'n," he said—but he neglected to key his mike and nobody heard him.

"We're like some goddamned taxi service," Spengler complained. "Pararescue gets all the glory."

I hesitated to respond. Spengler could be unnerving at times. Finally, I decided I couldn't let the remark go unanswered. "You've got to have a special kind of nerve to do what they do," I said, surveying the ground. "Guts."

"Did you say 'nuts?' You've got that right—you've got to be totally nuts to do what they do."

I'm surrounded by idiots, I thought.

We hovered and waited. The flames eventually died down, consuming all there was to burn, and the smoke thickened. Waves of birds continued to scatter, heading off in all directions.

"I hate all this waiting."

"Would you look at that," I said, watching a movement /'on the ground.

Spengler craned his neck to see. "Is that an elephant?"

"I believe it is," I confirmed.

"Goddamn. A fucking elephant. Look! There's another one."

We watched as the big gray mounds moved rapidly, rubbing against the trees, their trunks swaying, their whole bodies bouncing as they hastily made their retreat through the thickets of vegetation.

"There's four of them," Crazy Horse observed.

I first became aware of the change in wind speed while we were observing the elephants. It wasn't often that it happened. In fact, it was a rarity.

As a rule the air in Vietnam was so still it was stagnant. Rot was the rule rather than the exception and because of the lack of circulation, everything—food, clothes and especially bodies—could spoil in a day, leaving behind a putrid stench that saturated. It was everywhere all of the time. I was pretty damned sure Vietnam was the worst smelling country on earth.

But, sometimes the weather changed unexpectedly and the winds swept down from the highlands, whipping themselves into a frenzy as they made their way to the ocean. I and my team were directly in its path that day.

I watched the trees bend. The change was subtle at first—of no concern. But in the next instant, without presage, the leaves were vibrating so hard they were being torn from their branches.

Our cab was suddenly a compression chamber—the air painfully pressing against our ear drums as the wind whistled in one cargo door and out the other. The chopper resonated like a woman screaming.

Loose papers, candy wrappers and maps were whisked away or whipped from our hands. They whirled around like miniature tornadoes before being suddenly sucked out. Crazy Horse's sunglasses were torn from his face. He reacted fast, but he wasn't quick enough. They slid down his arm and tumbled over his fingertips before disappearing out the door.

"Fuck me," he cursed as he helplessly watched his shades catapult into space.

"There goes the goddamned flight log," Spengler yelled.

In the beginning it was just a matter of adjusting the controls to keep the ship steady. But, the wind came in great gusts, lasting as long as twenty or thirty second—long enough to send one chopper careening into another.

It became a battle for survival just to stay aloft, and another to keep from sliding into a neighboring aircraft.

Anything that wasn't nailed down was soon airlifted by nature. The air became thick with dry leaves, red dirt and insects. The wind storm was rapidly becoming a dust storm. The hills in the distance were soon obliterated by flying debris.

Our Huey suddenly lurched to the side, traveling dangerously close to the nearest aircraft. Spengler cussed and sweated as he attempted to steady her. Fearing his reactions weren't quick enough, I bridled the controls away and swiftly brought the craft back under control.

"Is everybody having fun?" I quipped.

"How can you joke at a time like this?" Spengler snapped. He was visibly shaken and angry with himself. "We almost got killed!"

"Ha!" I sallied, Spengler's anxiety only encouraging me. "I laugh at danger. I mock it. I spit in the eye of death."

"Watch out!" Spengler shouted, pointing out my window. The neighboring chopper was out of control and headed our way. I instinctively pulled up on the collective and let the ship slide harmlessly beneath us.

"Whew! That was fucking close," said Spengler.

The pilot finally got his ship under control and waved. "Good reflexes, Olstein," he radioed. "Thanks for saving my life."

"Thanks for not killing me," I replied.

"You're a gentleman and a scholar."

"Think nothing of it—less even," I retorted as Snagglepuss (cartoon character) might. The pilot chuckled and returned cautiously to his position.

"Alright, gentlemen," came the Wing Commander's voice. "I just received word. We're moving in.'"

"Great," Spengler moaned. "As if we don't have enough problems to deal with." Reluctantly the five ships pushed forward.

The winds remained treacherous and the gusts were downright terrifying at times. "I hate this. I really fucking hate this," Spengler complained.

"Follow my lead," the Wing Commander ordered. "Give yourselves plenty of room to breathe." Then as an afterthought, "and keep your speed down."

We kept our distance and flew at different altitudes for safety, drunkenly swaying while battling the winds as we made our way for the hills.

The tension in the cockpit was dense enough to materialize and assume its own identity. Spengler was gritting his teeth so hard his jaw was turning white. Even Crazy Horse looked ready to burst. His eyes bulged until it seemed they were about ready to pop from his head. "You know," I said in my Spokane backcountry drawl, "it's times like these that really make my anus pucker."

It was just enough to break the tension. Spengler sat back in his seat and chuckled; the gunners were obviously pleased. Crazy Horse rubbed his hands together briskly, squinting and shaking his head as his shoulders bounced up and down, laughing. "Gentlemen," I announced to my crew, "I think it's about time for a little mood music."

"Yes, sir!" Crazy Horse shouted, knowing what was coming next. He'd lost both of his top front teeth in a bar fight in Saigon a while back, but refused to see a dentist about it. When he spied his reflection in a mirror he thought he looked pretty damned good anyway—maybe better. The missing teeth only served to give him more character.

Flying with the "Cap'n" was one of the few things he enjoyed while serving in Nam. Captain Olstein was damned near as crazy as everybody else, except he had a unique way of hiding it. Rank helped. It also helped that he was smarter than most of his superiors and always kept them guessing. They were never quite sure what to make of him. But, what really endeared Benny to the troops was his insistence on the outrageous—especially when it came to music.

"I like my music loud," I said as I reached around and flipped a switch on the Roberts reel-to-reel. It was secured with web belts to the floor behind my seat.

"I hope you gentlemen appreciate fine music," I said, twisting in my seat and winking. The giant reels began turning, dragging the tape over the silver playback heads.

"Yes, sir. I sure do," said Crazy Horse.

Three speakers, two in the back facing forward and one on the floor facing the rear were tied down with ropes. Each weighed exactly sixty-two and half pounds. Benny learned that for sure one day when he hand-trucked them over to the mess hall and had them weighed on the meat scales.

The speakers began to hiss. Spengler watched the ampere gauge on the control panel dip as the amplifier began sucking energy from the aircraft's electrical system. He looked sidelong at Benny and simpered. This was going to be loud.

They actually felt the sound waves wash over them as the orchestra hit the downbeat. John Barry's "James Bond Theme" blared from the giant woofers and the familiar twang of the 007 guitar lick sprang from the tweeters.

Heads bopped to the beat and chills ran up and down their spines. Grins grew so wide they were impossible to wipe off, and my arm broke out in goose bumps.

My crew loved it and though the sky had darkened from the dust and it was difficult to tell where we were going, the tension lifted and an air of confidence settled in.

As we drew near our objective small arms fire could be heard even over the thumping of the chopper blades and bone-piercing music. On the ground it was chaos; one Huey was down, bent and twisted—smoke billowing from its tail. Another, a Cobra gunship, was still in flames. Orange tracers came in long spurts, streaking out of the jungle. So much for wiping out the enemy with a few bombing runs.

It was cramped quarters in the clearing, barely large enough for two choppers to land safely. I drew up and hovered, waiting my turn as the lead helicopter and another landed.

"Tonto!" I wisecracked to Crazy Horse, spinning the ship around and exposing the right cargo door. "Give 'em hell!"

"Yes, masked man," Crazy Horse responded, and opened with a volley that began tearing the jungle up. When he stopped to reload, I swiveled back and let the other gunner have a go at it.

Soldiers—Marine Infantry, Army Rangers, and a few Pararescue—made for the choppers, their bodies bent at the waist as if the blades of the flying machines would lop off their heads if they stood up.

Some just carried themselves, hoisting M-16's or lugging radios; some helped with stretchers; two, Benny saw, carried wounded on their backs—piggy-back rides to the rescue ships.

The James Bond music was fading; another song would soon follow. I reached back and kicked up the volume.

Now the lead chopper was lifting off, and the remaining soldiers were bolting for the other Huey. Wounded soldiers were helped onboard while the dead were heaved aboard.

"It's our turn," Spengler keyed, as the space opened up.

I maneuvered our craft into position as combat troops rushed to the open cargo doors. The skids hadn't touched ground before soldiers were already piling in. I twisted in my seat to watch just as a soldier took a hit and fell back. A hole the size of a baseball appeared on his chest and blood splattered the interior of the cab. He slumped, collapsing into the arms of the soldier behind who tossed him back into the chopper then climbed in after him.

They all jumped and some ducked as bullets skipped across the nose of my ship, like someone drumming on a metal bucket. Spengler would have ducked, but he was restrained by his harness. "Let's get the fuck out of here," he shouted.

The cab was full. Benny pulled up on the collective and lifted off. They'd already risen over a hundred feet and were heading east when he was tapped hard on the shoulder and motioned to something on the ground.

A soldier had just burst from the brush and was high-tailing it for the clearing.

He stopped, pulled a grenade from his belt and hurled it back at the jungle. As it exploded he stooped and pulled a limp soldier from the ground, hoisted him over his shoulder and began running again.

Benny watched as enemy fire followed him, kicking up dirt, coming dangerously close. With his free arm he waved, and though he couldn't swear to it, it seemed for an instant their eyes met.

There were no rescue ships on the ground—a fresh wave of gusts was preventing the remaining choppers from landing.

"It's one of those crazy Pararescue guys," Spengler piped.

"Throw down a ladder," I hollered, turning my ship. "And fire everything you've got into the bush. We're going back."

A rope ladder was quickly tossed out as Crazy Horse swung his M-60 around and opened fire. The ladder unfurled and swung crazily below them as the chopper wavered, staggering in mid-air, struggling against the mighty winds.

"Do you think he'll make it?" Spengler asked, his voice shaking.

"I hope to fuck he does," I replied. "We're risking our lives for nothing if he doesn't."

That's when the rocket whizzed past. Its sparks hit the windshield and bounced off like glistening ping-pong balls, leaving behind burnt streaks and a look of terror on Spengler's face.

"Kiss my ass!" Spengler shouted in fear. "We have to get the fuck out of here."

"In good time, my pretty," I replied in the voice of the Wicked Witch of the West—"all in good time."

The Huey dipped lower, swooping like a sparrow. Benny looked like he was actually having fun, and to Spengler, that in itself was a scary thought.

All at once they were caught in a tail wind and forced aloft, gaining altitude rapidly. Like being twiddled by an invisible hand, we were thrust about like a child's plaything. Then, just as quickly, the phantom lost interest and let go. The aircraft toppled on its side and dropped.

As the ground rushed up to meet us, Spengler felt his stomach lurch into his diaphragm and thought he was going to blackout. He watched Benny work feverishly, attempting to regain control, and thought, "this is it, boy. Goddamn if this isn't it. Shit." And, closing his eyes he hoped if he was going to die it would be mercifully quick and painless.

Then something happened. He didn't know what, but when he opened his eyes and looked, Benny was wearing a smile and to his relief, they weren't falling anymore. The craft had miraculously leveled out and they were gaining altitude once more. Benny winked at him and keyed, "welcome back." Spengler managed to return a grin, but he had been seriously shaken and failed to appreciate the humor.

All this time the rope ladder swayed in the wind, beckoning, baiting, luring the fleeing soldier. Sometimes the ladder ran along steadily, dragging on the ground ahead of him as if it weren't attached to anything, as though it had a life of its own. Just some crazy old rope ladder out for a stroll, skimming over the ground on its way to who knows where. Other times it hovered a foot or two above the ground, rushing through the grass, blazing a trail, twisting in the violent gusts and banging into anything in its path. He ran after it, picking up speed, digging his heels into the dirt for leverage, moving with the rhythm of a short-distance runner—running for his life.

He was almost there, reaching with his free hand—his fingertips just brushing one of the slats when the ladder was whipped away, snapping in the breeze. But, he was determined and never faltered, never slowed—confident that if the timing was right, all would be well. He had a hell of a lot more living to do and it wasn't about to end here—not now—not like this.

"I don't think that poor bastard is going to make it," said Spengler.

"He'll make it," said Benny.

Then, just as quickly as the rope was whisked away, it suddenly changed direction and slammed into the soldier. Not surprised, he caught hold. As if it was all a part of some fantastic dance routine, he stepped onto the bottom rung and was lifted skyward, rising swiftly into the air.

"He's on!" one of the soldiers shouted.

"Well, let's hope to fuck he can hang on until we get clear of this shit." I swung the chopper around and throttled up. I had to climb high enough for the ladder to clear the treetops—that part was easy— the challenge was battling the winds and hoping to hell I didn't get a surface to air missile fired up my ass in the process.

As soon as it was safe, I lowered the bird and turned the controls over to Spengler as I unsnapped my harness. Before the skids had a chance to touch ground, I jumped from the craft and waded through the tall grass that moved like ocean waves from the downwash.

"Thanks for coming back for me, Captain. You're a lifesaver," the soldier shouted over the racket of the whirling blades.

"Benny," I corrected him. I hated being called by my rank.

The soldier squinted through the wind and dust at my name tag. "Olstein."His eyes lit up and a smirk crossed his lips, "Benny?"

"That's right," I confirmed as I took one end of the wounded black soldier. "This guy's out cold."

"I think he'd dead."

I couldn't believe it. "You risked your life for a dead man?"

"Goddamned right I did. I'm not gonna let those fucks have him. He's ours." I nodded. The dead soldier was heavier than I expected.""Ever been to Seattle?" the soldier shouted.

"Yeah," I shouted back as we sidestepped through the grass and made our way to the chopper. "I went to high school there."

"Indeed."

I stopped and looked into the eyes of the Pararescuer. The wind from the blades whipped my shirt and tossed my hair, clogging my ears with that incessant staccato. Whop-whop-whop-whop. There was something about the soldier that was unsettlingly familiar. Whop-whop-whop-whop. "Stan?" . . .

. . . "You know, the goddess with the big tits in science class. I still fantasize about her." The words were far away, trying to break through.

Yep, here we were seven thousand miles away from home and who should I meet up with, but my old high school chum, Stan—now Pararescue Tech Sergeant Stanley Richards.

It had been several months since that unusually windy day when the delta winds got angry and swept down from the highlands.

"So—tell me about them," said Stanley, rousing Benny from his reverie.

They were sitting across from each other in an Air Force mess tent. I had just finished my breakfast; the latest letter from Nellie had been neatly folded and tucked away—but I could feel it through my pocket and was constantly aware of its presence.

"Tell you about them?" I wasn't sure what *them* he was talking about anymore.

"Yes, them. The eighth wonders of the world. Donna's tits."

Stanley suddenly rose to his feet and addressed the mess tent at large. "Could we have a moment of silence please—out of respect." He bowed his head and though nobody knew who he was referring to, they stopped eating and put down their forks, giving him his moment—most assuming they were paying homage to a fallen comrade. "Thank you," he said and resumed his seat.

I shook my head and smiled. "Ah, yes. Donna's breasts—definitely worthy of our deepest respect. A sight to behold, and certainly deserving of our undying devotion."

"Indeed. As far as I know, you're the only one who actually got a good look at them. I understand she forced them upon you—practically thrusting them in your face."

"Yes, she did indeed, my friend."

Stanley nodded and pursed his lips, thinking. "I should be so lucky. She could smother me with them and I'd die a happy man."

"Your mourners would weep with envy. Oh, by the way—" I pulled a paper from my pocket that looked like it had been wadded into a ball at one time. I smoothed out the wrinkles and placed it on the table before Stanley. "What the fuck is this?"

"A seriously damaged government document, I'd say."

"Read it."

"It looks official."

"Oh, it's official alright." I watched Stanley study it for a moment then push it back and resume eating.

"Looks like somebody recommended you for a Bronze Star." "Now, who the fuck do you suppose did that?"

Stanley shrugged and continued to eat.

"It was you, wasn't it?"

Stanley stopped eating and put down his fork. "You saved my life, Benny—it was the least I could do."

"Well, you can do me a favor by undoing it."

"What's the matter, don't you want anybody to know you're a hero?"

"I'm not a hero. I have a job to do and I do it. Period."

"Same here." Stanley shifted his eyes away from me and studied my plate. "I did what I thought was right, that's all. I'm grateful."

"I don't need a fucking Bronze Star. You're alive—that my reward."

"So, divorce Nellie and marry me, then."

"I'm serious, Stan."

"Okay, okay, okay. I'm sorry, okay? But, I'm not retracting the motherfucker and besides nobody would listen to me if I did. The nomination has been verified by two other soldiers already."

"Great," I sighed as I refolded the document and stuffed it back in my pocket. "Thanks a whole fucking lot."

"Hey, I said I'm sorry. I promise I'll never do it again." He grinned. "Now, let's get back to Donna's tits."

"Fuck Donna's tits."

"I'd love to."

"Do me a favor—don't do me any favors. People will treat me different now. I hate that. I don't want to be treated different. I like the way the commanding officers think they're better than everybody else and treat the rest of us like shit. I like knowing they're wrong and I like knowing that they haven't the slightest fucking idea that they're wrong. It gives me a reason to be. A raison d'être."

"Fine. Understood. Please accept my humblest of apologies. I will personally kick the ass of the next person who says anything nice about you, alright?" Stanley picked up his fork and resumed eating. "Are we still friends?"

"Barely."

"Good. Now describe them to me."

"Describe what?"

"Donna's tits. And, don't leave out a single breathtaking detail. What did they smell like?"

I laughed as I rose. "Why don't we talk about it tonight. I've got to get to my flight briefing."

"Okay, fine—but, don't forget. What am I saying—how could anyone forget those marvelous pechos?"

"Pechos?"

"Sí señor. You know what? I've always wanted to poke one of her nipples right here in my eye. Just rub it around for a little bit and polish my lens with it."

I laughed as I picked up my tray. "One of these days," Stanley continued, "when I get out of here, I'm going back to Seattle to find her. And when I do, I will camp out on her front porch until she grants me an audience so I can worship her. I know," he said, as if the idea had just occurred to him. "I'll build an alter to her," he hollered after my retreating figure . . .

. . . Yes, the day had started out good, great even, but somewhere along the line things went terribly wrong. A bullet had punctured the fuel tank and we were running on empty.

The blonde kid with part of his head missing hadn't budged. Cold air rushed in through the open cargo door and tossed his platinum hair, but he was turned away and I couldn't see his face. The black soldier who remained by his side was still talking to him and I wondered if he could hear.

The soldier that had been flailing remained still. It didn't make any difference if it was the morphine, or if he'd fallen unconscious—at least he was out of his misery for now—and if the medic didn't succeed in stopping that one artery from squirting like a fire hose he'd be out of his misery for good.

"We're losing altitude fast," the new copilot announced, his voice breaking.

"We'll make it," I said, looking over and reassuring him with a smile. "I'm taking us down. There's a clearing up ahead. Call in our coordinates. Tell them to send in the cavalry. And don't worry," I said as the copilot changed frequencies.

I pulled up and hovered at 1000 feet. We were directly over a rice paddy. Perfect, I thought, as we began our descent . . .

. . . Jeffery swam in his consciousness and wondered if maybe he'd been wounded worse than he thought. He was so cold. Then again, maybe this was all a dream—- maybe he was sleeping and he'd forgotten to close the window.

Maybe he was already dead. Maybe he was in purgatory, or better still, a soul in transit being transported back in time to relive the whole thing—right the wrongs, cross the t's, dot the i's, fix what was broken. That would be the fair thing.

But life wasn't fair. Never had been.

"Life may not be fair sometimes, but God is. You mark my words," Grandpa had said, *"when the time comes, He'll give ya a fair hearing. And, if deep down inside you're really a good boy, why it won't make for a hill of beans what you've done otherwise. Providing, of course, you haven't gone out and killed someone or tortured some poor animal to death."*

"But, what about soldiers," he'd asked.

"Soldiering? Soldiering's different. That's the only time killing's okay."

"Were you ever a soldier?"

"Sure I was. World War I."

"You ever kill anybody, Grandpa?"

"Me? Hell, no. I feel bad if I have to swat a fly. But, I'm a good bluffer. Bluffed my way out of many a tight spot. Bluffing's an art—a form of self-defense. And hell, if nobody gets hurt, it's a good thing."

But there was no way Jeff was going to bluff his way out of this. He couldn't move; he couldn't see. He could barely feel. At least he wasn't in pain. For that he was thankful.

Something told him he was falling. The little stones floating in his middle ear slammed against the roof of his semi-circular canals . . .

. . .That was when the engine whined down and quit altogether.

"We're going down," I heard myself say a little louder than I intended. Yep, the mission had started out pretty darn good. Everything had gone according to schedule; like clockwork. It was going to be an easy operation—and up to a point it was. Like items being checked off on a laundry list, everything flowed smoothly, precisely as planned—well, almost.

0600: Arrive at flight line.

The flight briefing had gone reasonably well. Major Quarrey, the ex-naval pilot, their usual wing commander, was in the infirmary. He complained of stomach pains during the night and early that morning his appendix burst. He was in surgery as they spoke. Benny, Captain Olstein to his superiors, was told he would be commanding the operation that morning.

"So, that's it. You'll take your orders from Captain Olstein," the squadron commander finished as the pilots refolded their maps. "This should be a simple operation. Just get in, drop your load and get out. Anybody got any questions?" There weren't any. "I'll expect you all back here by ten-hundred hours. Dismissed."

When I arrived at the flight line I found a squad of soldiers dutifully awaiting transport. Some of the older ones were milling anxiously, anticipating, dreading, chain-smoking. The scent of marijuana was heavy. A few dozed, or tried to.

Someone had brought a portable radio and several soldiers huddled around to listen to Everything Is Everything lend their voices to an ancient Commanche chant, "Witchi Tai To." Some of the boys were joking, but their laughter was nervous. Most were solemn, smoking cigarettes, staring at the tarmac or their combat boots while wind chimes melded with flutes and the moment became music.

The voice of Jimmy Pepper caressed the eardrums. His tenor sax floated out over the flight line riding upon the oppressive waves of heat all the way to my helicopter.

I couldn't wait to get my bird up. A higher altitude was the only place in this godforsaken land where the temperature was tolerable. The higher I flew, the cooler it got. Sometimes it got downright cold. The air whipped through the cargo doors, tossing my hair and violently blowing up my shirt, rippling—like sheets snapping on a clothesline on a windy day.

"Hi."

A tall and lanky soldier wearing a bright orange flight suit sauntered up. "Name's Henderson," he said, extending his hand. "Billy Henderson. I'm your new co-pilot."

"I kind of figured you might be," I said. "Do I detect an accent?"

"North Carolina. A little town east of Winston-Salem. One of the prettiest spots you'll ever lay eyes on. Got me a wife and two kids, twins, waitin' for me back home, too. Ever been there?"

"Nope. Here," I said, handing him the clip-board. "Why don't you help me with the safety check. There'll be six of us heading out this morning. We'll be flying command. Those are the other five slicks over there," I said, pointing farther down the line. "We'll be transporting the troops. Two cobra gunships will lead the way to soften up the landing area before we get there."

"Why's that?" Billy asked, fidgeting.

"LZ Charlie might be hot, so we're not taking any chances." I looked at my watch. "Another twenty minutes or so and we're out of here."

Billy nodded, taking it all in—plainly worried. "What happened to your last co-pilot, if you don't mind me asking?"

"Got his ear shot off."

"Oh."

"That flight suit your wearing," said Benny, watching Billy look down admiringly at his outfit. "Unless you want to be used for target practice by the NVA this morning, I suggest you head back to the barracks and change into something less conspicuous." Billy's face fell.

"Fatigue pants and a t-shirt'll work," said Benny. "That's probably your best bet." Billy looked like maybe his feelings had been hurt or he was embarrassed.

"Don't let it bother you. Everyone here made the same mistake the first time." Billy looked relieved when he heard those words. "That fellow coming up over there will be our gunner."

Billy turned in time to see a soldier, a Staff Sergeant by rank, at least a foot taller and twice as wide as himself, stroll up, a belt of M60 shells draped over his shoulder. "Kimo sabe," said the gunner, nodding to me. No military etiquette here, he'd worked with me before. He knew better. "Me bring extra belt. Now we have plenty ammo."

"Crazy Horse, meet our new Peter Pilot."

"It was nice knowing ya," Crazy Horse said with a nod. "By look of outfit me think maybe you won't be with us very long." He grinned wide, displaying his missing front teeth.

"He was just heading back to change."

"Ah," Crazy Horse said, heaving the ammo belt onto the bed of the Huey. "Me savvy."

"Don't worry, we won't take off without you," said Benny.

"Him plenty young warrior," said Crazy Horse, watching the new co-pilot hurry back to his hooch. "Me think maybe he might be too young." Sometimes Crazy Horse really loved talking like Tonto of the old Lone Ranger series, and sometimes it would go on for hours.

"That's right, Tonto," I said in my best Clayton Moore voice. "But I have a hunch he might turn out to be a pretty good soldier someday."

"Hmph. You may be right, kimo sabe. We wait see."

0645: Load troops.
0700: Lift-off.

Get in, then get the hell out, all in one piece if possible—those were my instructions, and that's what I was mulling over as I hit the ignition switch.

"This is my favorite part," I said over the whine of the engine as I secured my harnesses.

Billy, newly attired and less conspicuous, was too nervous to respond as the blades began turning overhead. With each pass, the blades whipped faster and the engine whined higher. Their thunder chattered its way to his very core; his insides vibrated until it seemed his organs would tear loose and float free.

"Coyote one-zero," came the flight controller.

"One-zero," I answered.

"You are cleared for lift-off."

"That's a Jolly Roger ten-four."

Crazy Horse beamed—he just loved me, the "Cap'n."

The skids lifted smoothly and floated over the tarmac as I turned my chopper in a 180 degree arc. They were waiting for my cue, the Hueys poised, their blades spinning, like hornets ready to pounce.

"Hold on to your balls, boys," I keyed. "Here we go." I throttled up, turned the chopper another 180 degrees and lifted off. "LZ Charlie," I said, "ready or not—here we come." My squadron followed.

I hit the playback switch on the tape deck. "And now," I said over the intercom, "for a little traveling music. Hit it boys!"

As if on cue, Balloon Farm blasted "A Question of Temperature" from the speakers, giving us a mighty send off, their growling voices screaming over the thump of the rotor blades.

When we reached a safe altitude Billy asked, "how long will it take us to get there?"

"Not long. All told, probably around twenty-five, thirty minutes."

Billy nodded and looked nervously out the window. At this height he could see the ocean way off to the east. He watched the ground and the changing scenery—areas cleared by the military gave way to jungle; twisting rivers gave way to areas cleared by bombing; patchworks of rice paddies and old men and women bent over working the fields, buffaloes pulling plows; clusters of mangroves, golden cypress, teak, coconut and rubber trees making patches in verdant hues.

Balloon Farm gave way to The Electric Prunes who were trying to get us to the world on time, followed by the Beatles who reminded us that our mothers should know, and finally the Chocolate Watch

Band talked about girls— yeah, yeah, girls, as LZ Charlie finally came into view.

"Thar she blows, maties!" I announced, turning off the tape player.

In a small clearing up ahead the damage left behind by the gunships was apparent.

0735: Drop your load.

Almost there, I thought. Land, let the troops get the fuck off, then get the hell out. Don't wave, don't salute, and don't say "so long, Charlie," just get the fuck out. That's the plan.

The Cobras had flown ahead and done their job softening up the landing zone. The area was peppered with craters and smoke was drifting up where some trees once stood. I wasn't sure if the gunships had made it safer to land or done the opposite by announcing our location.

A team of Green Berets stood by, a yellow smoke grenade marking their location. It was a changing of the guard—drop some off, pick some up. Even after a stretch in the field they didn't look all that ragged. Some wore shirts, some didn't, some smoked and paced. As the choppers drew nearer they gathered their equipment and tossed their cigarettes.

Benny came in low and fast, then pulled up and hovered. The rotary wings slapped the air as the troops piled out, disappearing into the waving grass.

The landing zone was surrounded by a thick curtain of jungle, the trees swirled in the downwash. The ground was wet like it had rained the night before, but the sky was clear now and the sun was hot.

"Looks like they might have wounded," Billy keyed, nodding in the direction of the soldiers.I agreed and lowered the skids to the ground to ease their loading.

The troops continued to pile out while the Special Forces Team stood their ground. Benny watched a black soldier talking to another Green Beret with platinum hair. His arms flew about and his face twisted comically as he exaggerated his story. Then they were both laughing and the black soldier rubbed the other's shoulder affectionately. They were obviously good friends. They'd probably been out in the bush for a couple of weeks and were damned glad to be going back in.

The next slick swooped in. "Right next to you, Commander," the pilot radioed.

"Commander?" I looked over and the other pilot waved, grinning. "Why you just call me Your Highness?"

"Sorry, my mistake. Your Highness."

I gave him the finger and the gesture was promptly returned. Troops were already dropping out of the neighboring Huey's cargo door.

I looked back. My chopper was nearly empty, only a couple of soldiers to go. I flipped on the tape deck and the sweet voices of the Shirells blared "Soldier Boy".

The music was good enough to dance to and that's just what a few of the Green Berets were doing. The blonde, in just his T-shirt and jungle pants was snapping his fingers and moving with the groove. The black soldier was dancing with him. Benny's head nodded to the beat and the Green Berets, seeing his chopper was nearly empty, advanced.

If I were to look back to pinpoint a moment in time when everything changed, it would have to be then—that one split second before the world came to an end—like the instant just before the big bang.

Another slick swooped in and hovered. "On the other side of you, O' Fearless Leader," the pilot keyed.

"That's right. You guys just keep it up," I acknowledged.

"Roger that. I always try to keep it up."

We made eye contact and exchanged middle-finger salutes.

Soon all six Hueys were hovering and dropping their loads. A few were taking on wounded.

With all the racket from the choppers and my homegrown concert, nobody heard the first shots. Soldiers were suddenly dropping for cover while others were dropping from wounds as a firefight broke out.

There was an explosion as a land mine was tripped. The Green Berets dispersed, trying to make smaller targets of themselves, zigzagging their way.

They'd been lying in wait. Somehow, the NVA knew in advance they were coming. Or, maybe not. Maybe the North Vietnamese Army had been off doing something else when the ruckus from the gunships caught their attention. Nobody would ever know for sure, and right now it really didn't make any difference. They were in trouble, and it was time to leave.

Billy nearly jumped out of his seat when the nose of their Huey was peppered with bullets. The soldiers were returning fire in every direction. They were surrounded, being attacked from all sides.

As the Green Berets rushed for the slicks they picked up wounded. Soldiers who only moments before had disembarked were now being tossed back into the very slicks they'd arrived on.

Two soldiers, the black and the blonde one, ran together. Without breaking stride, in one motion, they grabbed hold of a fallen soldier. The injured soldier seemed to glide effortlessly between them, the toes of his boots floating above the ground as they headed for my Huey.

All of a sudden the blonde stumbled and fell, disappearing into the tall grass. The black soldier ducked and hit the ground with a roll. A moment later he was scrambling back to his partner.

The pilot on my right waved as he began to lift off. "It's getting a little too warm around here. Heading back," the pilot radioed. That's exactly what I wished I was doing.

Another slick began lifting off. Then another.

Shit, I thought, I'm going to be the last one here. Come on, you fuckers, get on board for Christ's sake!

Just then, a barrage of B-40 rockets screamed out of the jungle. Most went wide and missed. But, two didn't, and the choppers on their receiving end, one just lifting off and the other still hovering, exploded in bright balls of orange fire along with their crews and passengers.

Billy looked like he was going into shock. This was not a good day for him—his first day in the field, his first mission. I was pretty sure if Billy had managed to find his voice it would have been a scream. Now, all he could do was sit frozen and stare in horror at the fiery debris and charred body parts falling to the ground.

The Green Berets were piling on board as bullets buzzed past, some hitting the ship. The black soldier was still in the tall grass, kneeling over his partner. The blonde had taken a hit.

A corpsman approached at a crouch and together they carried him to the waiting chopper. The blonde had a head wound. The black soldier climbed aboard and tended to his friend, unconcerned with the rest of the world or the soldier who'd just helped him—and gone back for more wounded. He caressed the blonde's forehead and spoke softly to him though I doubted he could hear any more.

The corpsman was back, this time with a soldier who's legs had been blown off at the knee. He was unconscious and coming around, his cries groggy at first, but growing louder. He was soon screaming and thrashing as the corpsman pulled him on board and attempted to pump him full of morphine and apply tourniquets.

That's it, everyone's on, I thought. The rest were engaged in the firefight, my chopper was full and it was time to save my own butt. I throttled up and lifted off, rolling to the south, barely gaining enough altitude to clear the jungle.

The skids of the UH-1 clipped the treetops, they were still too low. The rotor changed pitch. I tugged on the stick, gave it more gas to reduce the drag and maneuvered for more lift.

Below us now, the American soldiers and their enemy were invisible. Looking down, it was just another beautiful sunny day in South Vietnam as a cloud of birds burst from the mangroves and scattered.

The helicopter rolled off to the right, picking up speed as we rose. We'd taken several hits and were losing fuel fast. It wasn't long before our tank was empty. Fuck it—count your blessings; we were damned lucky we'd made it as far as we had. Hell, we were damned lucky we'd made it out at all.

"Coyote one-zero be advised we're taking fire from the south." It was the voice of a pilot from one of the choppers returning to base. Counting mine, there were only four now.

Just then, streaks of orange blipped past our open door and I instinctively ducked, though it wouldn't have done me any good.

"Holy Christ Almighty!" Billy screamed in terror.

"Tracers!" Crazy Horse shouted, manning the M-60. "Be advised we're taking fire right up our ass, Cap'n."

"Give 'em a piece of your mind, Crazy Horse," I keyed.

"I'll give 'em more than that." Crazy Horse gripped his weapon tight and tipped it high, aiming low. "Dirty rotten sons o' bitches," he swore through clenched teeth as he opened fire. The cabin resonated with the sudden concussions of rapid-fire machine gunning, overpowering the thumping of the rotor blades.

"Mayday! Mayday! We're going down!" Billy screamed into his mike."

"Calm down, Billy," I said, snapping the tape player off. "They know where we are. They're probably responding already."

But, Billy didn't hear me. All he heard was the silence from the engine that had stopped, and the whistling thump of the rotor as it idled down. I glimpsed over long enough to see my copilot clasp his hands together in earnest prayer. "Please God, please God, please God . . ."

As soon as the engine shut down I pushed hard, lowering the collective. This allowed my helicopter to drop to the ground more rapidly, which shocked Billy. He worried that I was insane or had a death wish. But, I knew the maneuver would conserve the remaining rotor RPM and prepare us for a softer landing later.

"Ladies and Gentlemen," I mumbled so only I could hear. "The Captain and crew want to thank you for flying with the U.S. Armed Forces this morning."

We had entered autorotation. I knew anything I did to reduce the shock of landing would be to our benefit. I had wounded on board and any sudden jarring might be enough to kill some of them.

"Please return your trays to the upright position," I continued. "We hope you enjoyed your trip and anything we did to make your flight more pleasant."

Once the engines failed, the main rotor still had plenty of inertia to spin under its own power. By manipulating the cyclic and tail rotor pedals I coaxed the Huey into a smooth and steady descent.

"Those of you feeling slightly ill," I said, "may want to empty the contents of your stomach at this time in the handy-dandy barf bags provided for your convenience."

This was going to be a pinpoint landing. One for the books! I pulled back hard on the collective to cushion our landing.

"Please wait until the aircraft has come to a complete stop before exiting," I ended.

The rotor began vibrating heavily, shaking our insides. Outside, the heat and the dust danced. Through the haze I could see a blue sky above a dense jungle and a rice paddy that was rising up to meet us. The skids touched ground and the Huey sunk into the soft, soggy earth.

"Now all we've got to do is sit and wait," I keyed.

And, we did. I laid my head back and closed my eyes and listened to the rotor wind down, finally idling to a stop. Now it was just the insects and the birds and the monkeys from the nearby jungle.

"Cap'n," Crazy Horse whispered hoarsely, "that was the most beautiful landing I ever saw."

"Thank you, Crazy Horse. Your appreciation is duly noted," I replied, unable to suppress a smile.

We sat quietly again, listening, until Billy broke the silence once more. "Well now," he said, unsnapping his harness, "I think I've had just about all the excitement I can handle for one day."

That Billy was sure a character I chuckled as I watched him doff his helmet and slap a Yankees baseball cap on his head.

"Yankees?" I scoffed facetiously.

"Goddamned right," said Billy indignantly. "I may be from North Carolina, but I know a good ball team when I see one." He tossed his harness straps aside and tried to cross his legs in the cramped compartment.

"I don't mind telling you," he said in his slow Southern drawl, "I totally and completely despise this fuckin' place. It scares the Livin' P. Jesus right out of me." His remark didn't require a response. Vietnam scared the hell out of everyone.

I was thinking how good it was to be on the ground again, and was looking forward to heading back to the base when the world exploded and the controls were ripped from my hands.

A rocket grenade fired from the nearby jungle hit our side. In the same instant, the nose pitched forward and what was left of our tail sailed skyward, flipping us over. Bodies and anything that isn't nailed down went flying. My world went black when the overhead canopy collapsed.

Later, when I would try to recall the series of events, my memory for the first time in many years would be hazy.

It was difficult to separate the real from the unreal. I slipped in and out of consciousness like a man drowning. When I thought I was just about to the surface and the muffled noises were just breaking through—I'd sink again, to be swallowed by the black.

Suddenly he was flying, rising rapidly like an ascending Ferris wheel. Then, unexpectedly, he dropped—fast and hard, landing on his back.

His world was no longer black. Light was seeping in ever so slowly through the haze. I'm not blind anymore, he thought. Maybe my injury isn't so bad, after all. Then he thought he heard music. It was far away but drawing nearer. Voices humming. Beautiful voices, like a choir, but now they were being replaced by the beat of wings flapping. Wings flapping? Yes. It couldn't be anything else—and they were coming nearer—some were already so close he was sure he could feel their wind.

He squinted against the blinding light. His world had turned white. Unsure if he could believe what he saw, he rubbed his eyes and tried to focus. I'll be damned, he thought, it's snowing.

I swam to the surface once more, my eyes fluttering open. It was hard to breathe. There was pressure on my chest; drawing a breath was painful and took great effort. I was suffocating. I tried to remember where I was and what I was doing there, but I was unable to come to grips before slipping under again.

It's snowing! Great, puffy, frozen flakes of white stuff was drifting to the ground. Snow. He wanted to get up and run and jump in it. And, he almost did when it dawned on him: Hey, this is crazy—it never snows in Vietnam. Then he heard Bo whisper in his ear, "Hell no, man. But, it sure as fuck can rain bullets though?" When he looked around, there was no Bo, he was gone, or he was never there, and his whispering voice was falling away. "I'll see you in a little while, good buddy. I'll be waitin' for ya."

"Waiting for me?" he asked. "Where?" But, there was no response and the sound of flapping grew louder. When he looked up he could watch the flakes of snow descend, growing larger as they neared the ground and his face. It was midday and the sun had no intention of setting. The sky was bright gray and the snow flakes continued to expand, growing to an enormous size and falling in his eyes where they melted.

I swam to the surface again. My head throbbed, something was pressing down upon me—heavy. My eyes fluttered open briefly then snapped closed of their own accord. But the memory of their vision remained. I thought I'd seen Jeffery—as close to me as the nose on my face. But, that was impossible. It must have been a dream.

Something was shimmering above, hovering. Something so bright he had to shield his eyes. He squinted to back the light.

"Jeffery." A voice was calling him. A man's voice, but a soft one. "Jeffery." The voice was near, now.

Finally, when he squinted hard enough, Jeffery saw the outline of his visitor. He was standing with his feet planted firmly apart, towering over him, his enormous wings arching heavenward, far above his head, while the lower tips brushed the surface of the new fallen snow, scattering it like dust. The wings never stopped moving. They pulsated, fluttering in a graceful cadence—like a butterfly at rest.

"Remember me, Jeffery?"

I must be delirious, imagining things, I thought, fighting my way to the surface again. That couldn't have been Jeffery. In my vision Jeffery was missing part of his face and while one eye was shut tight, the other was partially open, the turquoise of the orb clouded over and milky.

"Yes. Yes, I remember you. You're . . ."

"I'm your very own special angel," the voice whispered. He tried sitting up, and to his surprise, he could. It was like he'd never been wounded at all. He sank back and rested against a tree and smiled at his heavenly visitor. "I'm glad you came back."

I fought my way back once more, this time resisting the pressure that was suffocating me. In a burst of sudden strength, I pushed whatever it was away and as I did my hand brushed something hard. My fingers closed over it before losing consciousness again.

After a time, his body grew warm, no longer as cold as it once was. His visitor must have brought company with him because Jeffery could feel the warm breeze of their wings flapping against the frosty air.

"What are you doing here?"

"I've come to take you home."

"Home?"

"You're coming with me."

"Am I going to heaven?"

"You are."

"But what about my cousins and my brother Benny? Will we ever be together again?"

"Someday. But, come—we must hurry."

Somewhere, either in my head or nearby I heard voices. Vietnamese. Friend or foe? I felt my body being lifted and pulled. This wasn't a carefully orchestrated transfer of the wounded, this was with a complete disregard for my injuries. My right leg burned and my head throbbed as my body was jostled and tugged and as I struggled to break through the haze, my foot hit a switch.

HELP.

"You're worthy, Jeffery"

"But, am I . . ?"

"You've been a good boy, Jeffery. You're a very special boy, Mister Man," he heard the voice whisper as his giant wings wrapped about Jeffery and pulled him close. *"There's no one quite like you in all the land."*

The angel smiled as the wind pushed his golden hair. It billowed about his head like a halo. Hovering above were more angels, their white wings flapping gently in the winter sky.

HELP.

The littlest angel sensed Jeffery's eyes upon it and looked down, floating motionless in space, and waved. Jeffery waved back. "Hello," he whispered.

"Hello," it whispered in return.

And suddenly he was soaring, rising higher into the universe, ascending within the arms of his angel, safe within the folds of his feathers.

On the ride back, my leg throbbed over the morphine. Stanley was sitting next to me, chuckling.

"Man, when you get back you're gonna have to write a letter to John Lennon and thank him for saving your life. We didn't know for sure where the fuck you were, and then all of sudden we hear the Beatles screaming 'Help'!" He shook his head, still chuckling. "The NVA's were crawling all over your Huey like flies on shit, man. But, something, maybe your boots or who knows what, but something hit the playback button on your tape deck, and the volume must have been turned up, full blast, because we heard it nearly a thousand yards away clear as day. Shit, the ones who got away will probably be deaf for the rest of their lives."

Come to think of it, I thought, my ears were kind of ringing. I looked down at my leg. My trousers had been sliced open from my ankle to my hip. My thigh was wrapped in thick, heavy bandages.

"You're gonna be okay, Benny." I looked at Stanley. "You're not going to be doing any flying for a while. Maybe never again in Nam, but you're gonna be okay. They're going to try and save your leg." I nodded. The morphine was making me feel pretty goofy.

"You're being medivaced to Honolulu. You'll be transferred to a plane when we land. You're gonna need surgery and there's some great hospitals there. What the hell, weren't you gonna meet your honey there in a little while anyway? By the time she arrives you'll be up and around. You'll see."

I gradually became aware of my surroundings. I was in a Huey. The thump of the rotors was comforting. It was strange being a passenger instead of the pilot, but it felt good, too and I decided I kind of liked it.

I asked Stanley what he was going to do with the rest of his life and Stan said his plans hadn't changed. He was still going to camp out on Donna's front porch and predicted they'd marry someday and have a million kids. He was right, most of that did come to pass— except six kids was more accurate. I flew up to be best man at his wedding, and Nellie was a beautiful bridesmaid. Stanley and Donna were, even to their friend's surprise, very compatible. Any quarrels they had usually ended in raucous love-making.

When Stanley turned the tables and asked me what I was going to do when I got back to the States, an answer wasn't immediately forthcoming.

"Hey, I spilled my guys, now it's you turn."

"I can hardly talk," I slurred.

"Yeah, morphine does have that affect on some people."

"I don't know, "I finally answered. When I looked down my hand was in the shape of a fist. I must have held a fist for a long time, because when I opened my hand the joints felt like they needed oiling. In my palm was a silver dog tag. I brought it closer to my eyes and tried focusing, but the vibration of the chopper and the drugs didn't make it any easier.

"It's Jeffery's," said Stanley.

"Jeffery's?" I looked around, but there were only men from the pararescue team and one other soldier who was injured. Crazy Horse lifted his head and grinned, his missing front teeth making a gaping hole in his mouth. I grinned back. It was nice to see him.

"He didn't make it, Benny. He's in another chopper. You and your gunner are the only survivors."

"Billy?"

Stanley shook his head. "I don't think he was strapped in. He went right through the windshield."

Stanley had a tough time dealing with me after that. He'd never seen me cry before and he wasn't sure if it was Jeffery, who I'd been transporting without knowing it, or Billy, my rookie co-pilot, or even the black soldier who was probably killed by the NVA during those final moments that I wept for.

The morphine probably didn't help any either, but maybe it was good I got a chance to have a good cry before we landed. Stanley held me and rocked me and let the tears fall. There's no shame in being human.

After leaving Nam, flying over the Pacific on my way to the hospital in Honolulu, looking down through the breaks in the clouds at the endless waves of blue, I replayed Jeffery's life and made some promises to myself:

I would always be good to Nellie. If my parents objected to our marrying then they could just go fuck themselves.

I would always be good to my children. They would always come first and unlike Jeffery, they would always know how much they were loved, and I'd tell them so every day.

I would tell the world about what I had seen, and what I knew, and hoped my words would make a difference in someone else's life.

The End

www.ingramcontent.com/pod-product-compliance
Lightning Source LLC
Chambersburg PA
CBHW070902120626
46546CB00001B/105